336
Topics in Current Chemistry

For further volumes:
http://www.springer.com/series/128

Aims and Scope

The series Topics in Current Chemistry presents critical reviews of the present and future trends in modern chemical research. The scope of coverage includes all areas of chemical science including the interfaces with related disciplines such as biology, medicine and materials science.

The goal of each thematic volume is to give the non-specialist reader, whether at the university or in industry, a comprehensive overview of an area where new insights are emerging that are of interest to larger scientific audience.

Thus each review within the volume critically surveys one aspect of that topic and places it within the context of the volume as a whole. The most significant developments of the last 5 to 10 years should be presented. A description of the laboratory procedures involved is often useful to the reader. The coverage should not be exhaustive in data, but should rather be conceptual, concentrating on the methodological thinking that will allow the non-specialist reader to understand the information presented.

Discussion of possible future research directions in the area is welcome.

Review articles for the individual volumes are invited by the volume editors.

Readership: research chemists at universities or in industry, graduate students.

Nelson L.S. Tang • Terence Poon
Editors

Chemical Diagnostics

From Bench to Bedside

With contributions by

C.H. Borchers · P.E. Brennan · A.G. Camenzind · A.G. Chambers · K.C.A. Chan · S.-C. Chiang · Y.-H. Chien · H. Cui · D. Domanski · D.S. Froese · L.-W. Hsu · P. Hui · W.-L. Hwu · H. Kambara · N.-C. Lee · E.W.Y. Ng · C.E. Parker · A.J. Percy · T.C.W. Poon · M. Shirai · D.S. Smith · T. Taniguchi · S.-F. Wang · L.-J.C. Wong · M.Y.M. Wong · W.W. Yue · W. Zhang

Springer

Editors
Nelson L.S. Tang
Department of Chemical Pathology and
Laboratory of Genetics of
Disease Susceptibility
Li Ka Shing Institute of Health Sciences
The Chinese University of Hong Kong
and
Functional Genomics and Biostatistical
Computing Laboratory
CUHK-Shenzhen Research Institute
Shenzhen, People's Republic of China

Terence Poon
Department of Paediatrics and
Proteomics Laboratory
Li Ka Shing Institute of Health Sciences
The Chinese University of Hong Kong
People's Republic of China

ISSN 0340-1022 ISSN 1436-5049 (electronic)
ISBN 978-3-662-50834-3 ISBN 978-3-642-39942-8 (eBook)
DOI 10.1007/978-3-642-39942-8
Springer Heidelberg New York Dordrecht London

Printed on acid-free paper

Springer is part of Springer Science+Business Media (www.springer.com)

Preface

In recent years we have witnessed a fast evolution of diagnostic technologies and their applications. The latest breakthroughs include a paradigm shift in sequencing and advances in mass spectrometry.

In 2012 we celebrated the 50th anniversary of the Nobel Prize in Physiology or Medicine for the discovery of the structure of DNA by Francis Crick, James Watson and Maurice Wilkins. This anniversary coincided with the complete decoding of the three billion base pair genome of James Watson, an effort which represented an unprecedented display of genomic diagnostics that just a few years ago was only possible in science fiction. This early success of complete genome sequencing of a few individuals was followed by the "1000 Genome Project" which aimed to decode comprehensively the base pairs of 1,000 individuals sampled around the world. The decoding of a significant percentage of the genome of an individual marks the new era of Personalized Genomic Medicine. Today's technology of massive parallel sequencing (also known as next generation sequencing and high throughput sequencing) has been made possible by using at least three Nobel Prize winning technologies, namely (1) sequencing of nucleic acids (Walter Gilbert and Frederick Sanger, 1980 Chemistry), (2) polymerase chain reactions (Kary Mullis, 1993 Chemistry), and (3) imaging semiconductor circuits (Willard Boyle and George Smith, 2009 Physics). More state-of-the-art technology is being incorporated into forthcoming generations of equipment in order to achieve a further increase in sequencing throughput and precision.

In this volume, a review of the existing and emerging massive parallel sequencing platforms sets the stage for a subsequent discussion of the application of genomic diagnosis (see chapter "Next Generation Sequencing: Chemistry, Technology and Applications"). This rapidly evolving and powerful technology moves quickly from the explorative stage to clinical application, despite ongoing discussions of various ethical, social, and regulation issues. The innovative utility of simultaneously analyzing a group of genes for making genetic diagnoses is reviewed in the chapter "Application of Next Generation Sequencing to Molecular Diagnosis of Inherited Diseases." One of the many applauded applications is the non-invasive prenatal diagnosis for genomic abnormality of the fetus. Screening for Down syndrome is now possible using a sample of peripheral blood collected from the pregnant mother. This is covered comprehensively in the chapter "Clinical Applications of the Latest Molecular Diagnostics in Noninvasive Prenatal Diagnosis." However, new analytic and bioinformatic algorithms are required to handle the vast amount of data or genetic variants generated by massive parallel sequencing. The chapter "The Role of Protein Structural Analysis in the Next Generation Sequencing Era" provides a review on how the current knowledge of protein structure and sequencing information help in the data processing.

The chapter "Emerging Applications of Single-Cell Diagnostics" introduces the emerging diagnostic area of single cell analysis. While all current diagnostic techniques sample hundreds or thousands of cells for analysis and return either a summative or average readout of an analyte in these hundreds or thousands of cells, no information is known about the concentrations or their variation in an individual cell. Therefore, there is a need to carry out analysis at the single cell level and it is hoped that this will lead to further development in the future.

Mass spectrometry has played a key role in metabolomics diagnostics in the clinics, allowing unambiguous identification of metabolites and their isoforms. Quantification at high precision can be achieved through various approaches. The application of multiple reaction monitoring in simultaneous assays of multiple analytes is covered in the chapter "Mass Spectrometry in High-Throughput Clinical Biomarker Assays: Multiple Reaction Monitoring." It is followed by a critical appraisal of the use of matrix-assisted laser desorption (MALDI) time-of-flight mass spectrometry (TOF-MS), especially surface-enhanced laser desorption/ionization (SELDI) TOF-MS, which possesses both the ability to discover novel biomarkers and quantification of known proteins and biomolecules (see the chapter "Advances in MALDI Mass Spectrometry in Clinical Diagnostic Applications"). The volume is concluded with the case of successful applications in the medical field. By using few drops of blood, newborn screening can identify babies with various genetic diseases soon after birth. These developments in tandem mass spectrometry methods in newborn screening are reviewed in the chapter "Application of Mass Spectrometry in Newborn Screening: About Both Small Molecular Diseases and Lysosomal Storage Diseases."

Hong Kong SAR, People's Republic of China Nelson L.S. Tang

NT received grant support from NSFC (31171213) and Shenzhen Municipal Government (GJHS20120702105523299).

Contents

Next Generation Sequencing: Chemistry, Technology and Applications .. 1
Pei Hui

Application of Next Generation Sequencing to Molecular Diagnosis of Inherited Diseases 19
Wei Zhang, Hong Cui, and Lee-Jun C. Wong

Clinical Applications of the Latest Molecular Diagnostics in Noninvasive Prenatal Diagnosis 47
K.C. Allen Chan

The Role of Protein Structural Analysis in the Next Generation Sequencing Era 67
Wyatt W. Yue, D. Sean Froese, and Paul E. Brennan

Emerging Applications of Single-Cell Diagnostics 99
M. Shirai, T. Taniguchi, and H. Kambara

Mass Spectrometry in High-Throughput Clinical Biomarker Assays: Multiple Reaction Monitoring 117
Carol E. Parker, Dominik Domanski, Andrew J. Percy, Andrew G. Chambers, Alexander G. Camenzind, Derek S. Smith, and Christoph H. Borchers

Advances in MALDI Mass Spectrometry in Clinical Diagnostic Applications 139
Eddy W.Y. Ng, Melody Y.M. Wong, and Terence C.W. Poon

Application of Mass Spectrometry in Newborn Screening: About Both Small Molecular Diseases and Lysosomal Storage Diseases 177
Wuh-Liang Hwu, Yin-Hsiu Chien, Ni-Chung Lee, Shiao-Fang Wang, Shu-Chuan Chiang, and Li-Wen Hsu

Index 197

Top Curr Chem (2014) 336: 1–18
DOI: 10.1007/128_2012_329

Published online: 31 May 2012

Next Generation Sequencing: Chemistry, Technology and Applications

Pei Hui

Abstract High-throughput next generation sequencing (NGS) has been quickly adapted into many aspects of biomedical research and begun to engage with the clinical practice. The latter aspect will enable the application of genomic knowledge into clinical practice in this and next decades and will profoundly change the diagnosis, prognosis and treatment of many human diseases. It will further demand both philosophical and medical curriculum reforms in the training of our future physicians. However, significant huddles need to be overcome before an ultimate application of NGS in genomic medicine can be practical and fruitful.

Keywords Next generation sequencing · Genomic medicine

Contents

1 Introduction 2
2 Chemistry 2
3 General Workflow of NGS 3
4 Evolution of Sequencing Technology 4
4.1 First Generation Sequencing 4
4.2 Second Generation Sequencing 5
4.3 Third Generation Sequencing 6
4.4 Emerging Next Generation Sequencing Technology 8
5 Common Problems with NGS Data 9
6 Applications 9
6.1 Whole-Genome Sequencing 11
6.2 Targeted Sequencing 11
6.3 Epigenetic Applications 12
6.4 Transcriptome Analysis 12
6.5 Microbial Population Analysis 13

P. Hui (✉)
Department of Pathology, Yale University School of Medicine, 310 Cedar Street, New Haven, CT 06520-8023, USA
e-mail: pei.hui@yale.edu

7 Limitations of NGS in Clinical Practice 13
8 Summary 14
References 14

1 Introduction

The conclusion of the human genome sequencing project in 2003 established the molecular basis for the understanding of many disease processes at genetic level [1]. As a result, the availability of reference sequence of the human genome has fueled the emergence of a new era of genomic medicine [2–5]. Advance in technology of high throughput next generation sequencing (NGS), also known as massively parallel or multiplex cyclic sequencing, is the key element that will enable the application of genomic knowledge into clinical practice. However, DNA sequencing and related genomic informatics must become more economical, informative, and readily applicable for the ultimate transition from empirical practice to precision medicine [6, 7]. The current speed of evolution of NGS technology is stunning and will soon result in the delivery of low-cost, high-throughput, and even portable DNA sequencing apparatuses to clinical laboratories. In fact, NGS has already begun to produce clinical benefits in some healthcare setting [8–10]. In the next few decades, genomic medicine driven by NGS will profoundly change the diagnosis, prognosis, and therapy of human diseases. It will demand both philosophical changes and curriculum reform in the training of our future physicians as well.

In order to get there, there are obstacles remaining to be overcome, such as developing sophisticated bioinformatics and computational biology techniques for analyzing vast amount of sequencing data, understanding variations of the genome, understanding genetic and non-genetic bases of human diseases, establishing effective ways to deliver evidence-based genomic medicine, and, finally, resolving ethical and legal issues in the practice of genomic medicine. In this chapter, the chemistry and technological background of NGS will be presented. It will be concluded by the direction of future technological development in these aspects.

2 Chemistry

DNA sequencing undertaken by the Human Genomic Project was completed in 2003 almost exclusively by Sanger's method, the first generation sequencing. In 2007, DNA sequencing was taken to the next level when the Illumina genome analyzer was introduced, heralding the era of next generation sequencing. Within 1 year, NGS was used successfully to sequence the first individual human genome (James Watson) in 2008 [11]. Now NGS technology is evolving at an unprecedented pace along with diminishing cost. It is expected that the cost will be approaching less than

$10,000 per human diploid genome in the coming years [12]. At the time of writing, a typical platform could produce up to 600 giga-base data in a sequencing run that lasts for 7–10 days. The data represent about 6,000,000,000 sequencing reads with a length of 100 bases.

Generally speaking, NGS employs DNA synthesis or ligation chemistry (sequencing-by-synthesis) to read through many independent DNA templates at the same time in a highly parallel fashion to produce a tremendous quantity of DNA sequence data. Sequencing-by-synthesis strategies [13] include a single molecule approach or ensemble approach (sequencing of multiple clonally amplified DNA targets on isolated surfaces or beads). Both approaches can be accomplished in either real-time fashion (DNA polymerase synthesizes without interruption) or synchronous-controlled fashion (DNA polymerase synthesizes in "stop-and-go" through controlled delivery of nucleotides or temporarily limiting extension using modified nucleotides or metal catalysts). The detection of signal can be achieved by fluorescent labeling of nucleotides, enzyme-coupled chemiluminescence assays for pyrophosphate, and pH change as result of proton release during each nucleotide incorporation.

One sharp contrast to the first generation Sanger sequencing is that NGS generates short reads of frequently less than 500 bp as opposed to over 1,000 bp. However, the massive depth of coverage, i.e. multiple reads over the same template DNA region, compensates for the limitations of short reads. NGS technologies have drastically increased the speed and throughput capacities over Sanger sequencing while reducing cost, even as we write. NGS may be classified into second and third generations according to their years of availability and chemistry. Second generation sequencing essentially uses DNA synthesis chemistry as employed by the traditional Sanger's sequencing. Third generation sequencing (Ion Torrent of Life Technologies, Inc and single polymerase sequencing platforms of PacBioRS, Inc) employs distinct chemistries, which will be elaborated in the following technology section.

3 General Workflow of NGS

Regardless of various sequencing chemistries, both second and third generations of NGS require highly complex pre-sequencing target preparation procedures and post-sequencing bioinformatics data analysis (Fig. 1). The pre-sequencing step includes target DNA enrichment and NGS library preparation. Target enrichment can be accomplished by amplification methods (PCR, Long-ranger PCR, or Raindance fluidigm PCR) or hybridization capture methods (by solid phase or in solution). NGS library preparation generally involves (1) fragmentation of the enriched target DNA by physical methods (sonication, acoustic wave or nebulization) into generally a length of 150–500 base pairs (library sequences), (2) ligation of the fragments to adaptor primers, and (3) clonal amplification of the library by either emulsion bead PCR or surface cluster amplification.

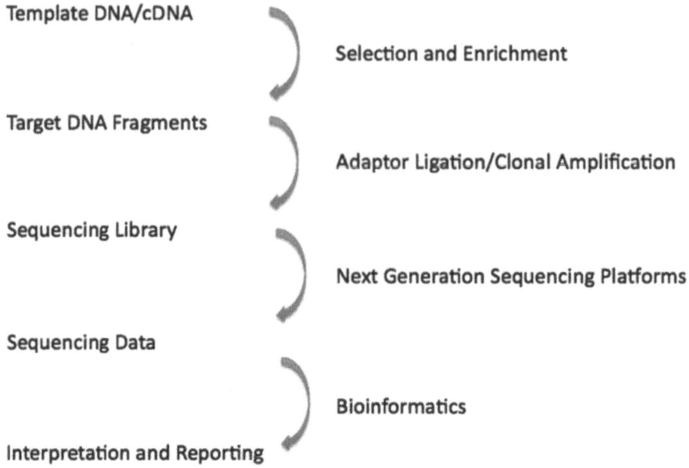

Fig. 1 Workflow of next-generation sequencing. This target enrichment and library sequencing approach are typically used in second and third generation sequencing approaches

The performance of sequencing reaction and capture of sequence data are the subject of a subsequent section. After a sequencing reaction, billions of reads are generated. Each read contains the sequence, typically ~100 bases in length, of a single template clone. Post-sequencing bioinformatics analysis generally involves sequence image processing to generate base sequences, sequence file conversion to readable files, and sequence alignment with reference DNA sequence for final variant identification and annotation. Adequacy of NGS relies on the sequence coverage and the depth. Sufficient coverage of DNA regions of interest is essential and sufficient depth of coverage (how many reads of the same region) is critical for accuracy and interpretation. Some common problems associated with NGS include sequencing reads are too short, resulting in difficulties in final sequence assembly or mapping; not all sequences are equally processed at high GC rich regions and homopolymers, amplification bias is inherent to some target enrichment processes, and sequencing errors (particularly longer reads) occur from 0.01 to 16 per 100 base read [14].

4 Evolution of Sequencing Technology

4.1 First Generation Sequencing

First generation sequencing technologies include sequencing by synthesis developed by Sanger [15] and sequencing by cleavage pioneered by Maxam and Gilbert [16]. Sanger sequencing dominated the biomedical research field before 2008. Standard four-color fluorescent labeling, where each color relates to one of the four DNA bases, has been the method of choice for detection by automated capillary electrophoresis (CE) platforms, commercially available from Applied Biosystems Inc., Life Technologies Inc., and Beckman Coulter Inc. The first complete human diploid genome (Craig Venter) was sequenced by Sanger's method in 2007 [17]. Although sequencing tasks in large comprehensive research projects have now shifted to NGS platforms, Sanger sequencing-CE platform will likely remain in significant use for targeted sequencing projects (biomarker identification and pathway analysis) and clinical diagnostic applications until small-scale NGS platforms become cheap and fast enough; this is an rapidly evolving area of industrial development (see below).

4.2 Second Generation Sequencing

Second generation sequencing is represented by Roche 454 pyrosequencing, reversible terminator sequencing by Illumina, sequencing by ligation of ABI/SoLiD, and single-molecule sequencing by Helicos. Using DNA polymerase or DNA ligase as their core chemistry, these platforms provide significant performance in large comprehensive whole genome sequencing projects [18]. Roche454 uses emulsion PCR to achieve clonal amplification of target sequence. The sequencing machine contains many picoliter-volume wells each containing a single bead and sequencing enzymes. Pyrosequencing uses luciferase to generate light for detection of the individual nucleotide incorporated into the nascent DNA [19–21]. Illumina (Solexa) uses cluster target sequence amplification on solid surface (bridge amplification). Sequencing is performed by adding four types of nucleotides, each labeled by one of four fluorophores and containing a 3′ reversible terminator. In contrast to pyrosequencing, DNA can only be extended one nucleotide at a time in the Illumina approach. After a fluorescent image of the incorporated nucleotide is recorded, the fluorophore along with the 3′ reversible terminator is chemically removed from the DNA molecule, allowing the next cycle to occur [12, 22].

Applied Biosystem/Life Technologies' SOLiD technology employs ligation reaction for sequencing using a repertoire of all possible oligonucleotides of a fixed length that are labeled according to the sequence position. Oligonucleotides are ligated after annealing. The preferential ligation by DNA ligase for matching sequences records the nucleotide position. DNA is clonally amplified by emulsion PCR on beads, leading to each bead containing only copies of the same DNA molecule. The beads are deposited on a glass slide [23] and sequenced. The sequences in terms of quantities and lengths are comparable to Illumina sequencing [20, 24].

HeliScope sequencer employs "true single molecule sequencing" technology [25, 26]. DNA fragments along with added polyA tail adapters are attached to the flow cell surface, followed by extension-based sequencing with cyclic washes of the flow cell with fluorescently labeled nucleotides similar to Sanger sequencing. Although the reads are short, recent improvement of the methodology provides enhanced accuracy of reading through homopolymers (stretches of one type of nucleotides) and also allows for RNA sequencing [18, 26, 27].

The second generation sequencing platforms vary significantly with regard to their throughput, read-length, and operating cost [12]. They are generally high throughput but highly expensive machines of scales of 0.5–1.0 million US dollars. The signal recording method is either fluorescence labeling or pyrophosphate chemical conversion, both requiring optical detection. The second generation sequencing platforms, although being successful in many research applications, suffer variably from high cost of instrument, complexities of sample preparation and chemistry (fluorescent labeling and enzyme-substrate reaction), complexities of optics and instrumentation, and read-length limitations [13].

4.3 Third Generation Sequencing

Sequencing technology evolves with the high demand for a low cost of technology. In line with the ultimate target goal of $1,000 per genome aimed at by the NIH/NHGRI invited grant challenge in 2004 for developing novel technologies, the third generation sequencing platforms are characterized by new chemistry, less operation time, desktop design, and lower operation cost. At the time of writing, three leading third-generation sequencers have emerged, which include Pacific Biosciences' real-time single molecule sequencing (PacBioRS), Compete Genomics' combined pro-anchor hybridization and ligation (cPAL), and Ion Torrent of Life Technologies, Inc.

PacBioRS is a real-time single molecule-single polymerase sequencing platform that can produce 1,000 bp read. Each chip has so-called zero-mode wave guided (ZMW) nanostructures of 100-nm holes, inside which DNA polymerase performs sequencing by synthesis with phospholinked nucleotides labeled with fluorophores which are introduced sequentially (Fig. 2) [28–32]. In addition to producing DNA sequence, monitoring the kinetics of nucleotide incorporation may help in the future to extract epigenetic information (e.g., methylation pattern) of the native DNA strands [33]. The platform has the ability to sequence mRNA by replacing DNA polymerase with ribosome [34]. The instrument of such configuration will, however, be expensive.

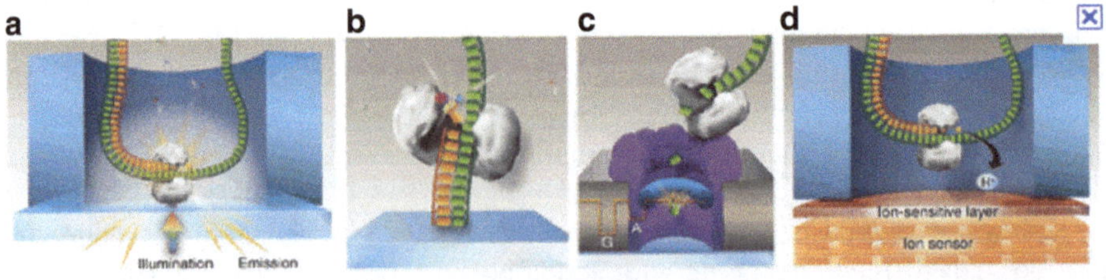

Fig. 2 PacBioRS real-time single molecular-single polymerase sequencing single stranded DNA template is sequenced by synthesis in nanostructure hole. (Copyright permission obtained from PacBioRS) [32]

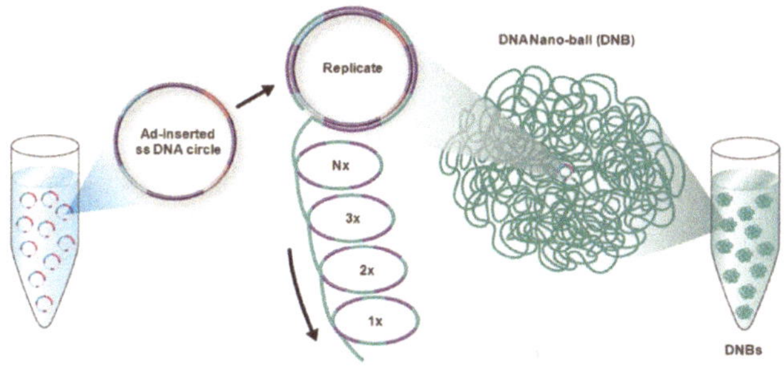

Fig. 3 Complete Genomics' Nanoball formation after rolling cycle amplification. These DNB are sequenced by ligation (©2010 Complete Genomics, Inc. Used with permission)

Complete Genomics announced a combinatorial approach of probe-anchor hybridization and ligation (cPAL) sequencing with the claimed highest throughput among third generation sequencers [35] (Fig. 3). The method uses rolling circle amplification of small DNA sequences into so-called nanoballs. Unchained sequencing by ligation is then used to determine the nucleotide sequence [35]. This method permits large numbers of DNA nanoballs to be sequenced per run and at low consumable costs [36]. The platform has been successfully used in clinical genome sequencing applications such as whole genome sequencing of individuals [37, 38]. However, mapping the short sequencing reads to a reference genomic database can be difficult, especially in the analysis of tumor DNA.

Ion Torrent technology (Life Technologies, Inc), perhaps the current most versatile and low cost method, has been delivered in the form of a personal genomic machine (PGM) as a benchtop instrument to research and clinical laboratories [39]. The sequencing chemistry of Ion Torrent technology involves proton release during each nucleotide incorporation by DNA polymerase. The dense microarray of individual microwell allows DNA polymerase to act on clonally amplified target DNA fragments. Beneath each microwell the Ion-Sensitive Field Effect Transistor (ISFET) detects the pH change as a result of each proton release and a potential change (ΔV) is recorded as direct measurement of nucleotide incorporation events (Fig. 4). The system does not require nucleotide labeling and no optical detection is involved. Ion-Torrent's PGM costs less than 100 K with sequencing capability adequate for small-scaled research projects or clinical diagnostic laboratories. Its introduction into the market hails the beginning of NGS as a commodity for biomedical and clinical applications. As a common feature to many other systems, it has multiplex bar-coding adaptors which allow simultaneous testing of multiple samples. The available chip sizes (314–318) capture 10–1,000 MB of sequence information per run. Although the current Ion Torrent operation is labor intensive, automation with one-step library preparation (one-touch sequencing library preparation kit) has recently become available to simplify the process. The limitations include short read-length (100–200 bp) and technical difficulties in reading through highly repetitive sequences and homopolymers, for which improvement has recently been made.

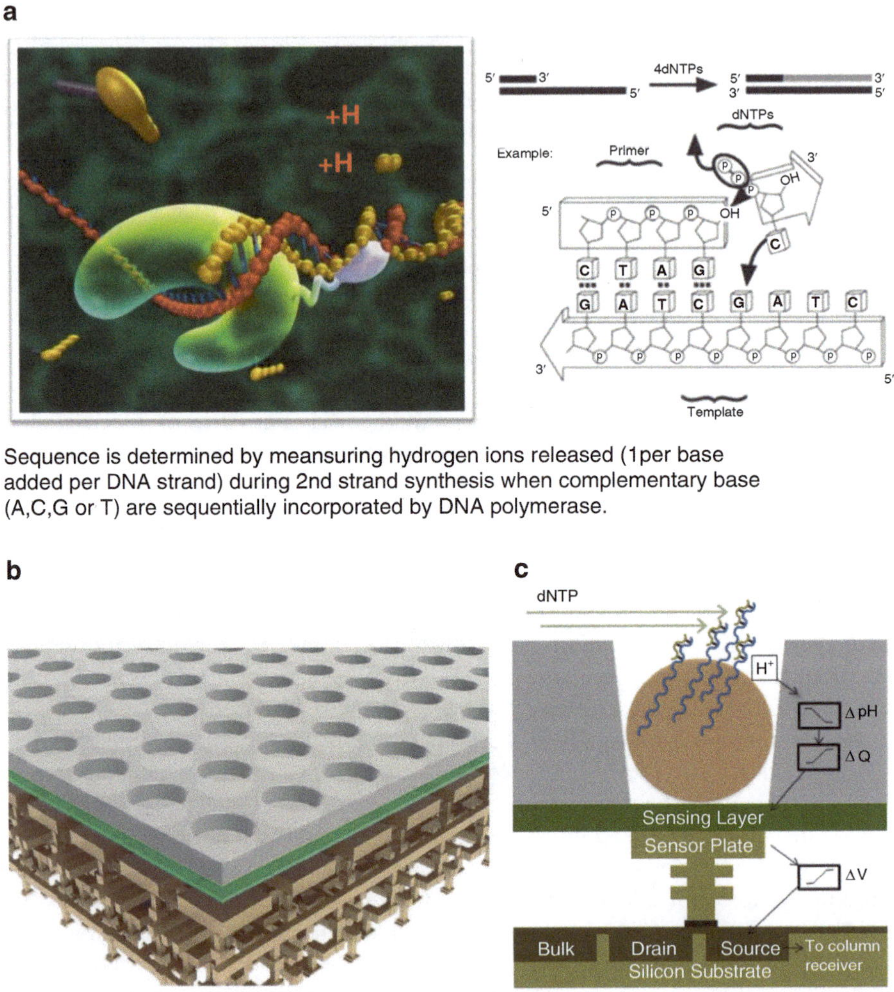

Fig. 4 Ion Torrent Technology. (**a**) Proton release when nucleotide is incorporated by the DNA polymerase into the DNA chain. (**b**) The Ion Torrent proprietary microchip design. (**c**) Cross-section view of a single well that houses ion sphere particles with a clonal amplified DNA template. A hydrogen ion (proton) is released when a nucleotide is incorporated by DNA polymerase. The proton is then detected by the sensing layer due to the change of pH, therefore translating the chemical signal to a digital input. (Copyright permission obtained from Life Technologies, Inc. 2011)

4.4 Emerging Next Generation Sequencing Technology

Oxford's nanopore technology has a different sequencing approach currently in the developmental phase. It uses the scanning tunneling electron microscope (TEM) that measures alterations of conductivity across a nanopore while a single DNA molecular is passing through. The amount of current that can pass through the nanopore at any given moment varies depending on the shape, size, and length of the nucleotide blocking the ion flow through the pore. The change in current through the nanopore as the DNA molecule passes through represents a direct reading of the DNA sequence. An exonuclease enzyme is used to cleave individual nucleotide molecules from the DNA, and when coupled to an appropriate detection system these nucleotides could be identified in the correct order [40]. Oxford's nanopore technology may also be suitable for integration into a system for analyzing epigenetic modifications. The α-hemolysin nanopore is a promising sensor for ultra-rapid sequencing of DNA strands within nanopores, which may provide additional sequence information using two recognition sites rather than one [41]. Furthermore, nanopore technology is free from some of the drawbacks of other platforms through elimination of the need for optical detection and DNA synthesis and even target DNA amplifications [40, 42–44].

5 Common Problems with NGS Data

There are some common technical problems associated with various NGS platforms. Short reads in many NGS systems result in difficulties with assembling and mapping to the reference sequences, particularly at repetitive regions. Not all sequences are equally processed and sequenced, and DNA regions enriched with GC content are particularly prone to low coverage. For NGS platforms with target amplification or enrichment, amplification bias may be introduced. Last but not least, sequencing errors are present essentially in all NGS platforms. Longer reads are prone to have error readings, particularly towards the ends. Repetitive sequences and homopolymers are also of concern for some third generation sequencers; however, rapid improvement has been made in recent months to overcome these problems. Increase of coverage and deep sequencing are important to correct some of these problems. Table 1 provides a summary of key characteristics of the current NGS platforms.

6 Applications

Genomic medicine driven by the latest development of NGS technologies will have profound impacts on our understanding of the pathogenesis of human disease and many aspects of clinical practice in the future: diagnostics, prognostics, and therapeutics. These clinical applications can be roughly divided according to defined target sequences: whole genome sequencing, targeted sequencing of exomes (selected or whole) or selected genes related to a specific disorder or category of disease, epigenetic mapping, transcriptome sequencing, and microbial population sequencing.

Table 1 Characteristics of current platforms of NGS

NGS	Platform (company)	Chemistry	Read length	Run time [14]	Through-put	Reagent cost/MB [14]	Minimal cost per run ($)	Advantage and limitations
Second	Roche 454	Phosphokinase fluorescent nucleotides	200–400	10 h	High	7–22	1,000–2,000	Long reads, difficulty in homopolymer reading, and expensive
Second	Illumina	Reverse terminator	150	1–14 days	Very high	0.1–0.7	1,000–3,500	Expensive
Second	ABI/SOLiD	Ligation	25–35	8 days	Very high	0.07–0.11	2,000–2,500	Low reagent cost, long turn-around-time, and expensive instrument
Second	Helicos	Single molecular sequencing	25–30	Unknown	High	Unknown	1,100	Expensive instrument
Third	Ion Torrent	Proton/pH detection	100–200	2 h	Moderate	0.93 (318 chip)-50 (314 chip)	750–1,200	Inexpensive, up to 16 bar-coding, Difficulty in homopolymer reading
Third	Complete Genomics	Probe-anchor capture and ligation	10	Unknown	Very high	Unknown	Unknown	Labor intensive, instrument not available in market
Third	Pacific Bio	Real-time single polymerase	1,000–2,000	0.5–2 h	Unknown	11–180	Unknown	Greater base read mistake, instrument not available on market

N.A. Not available

6.1 Whole-Genome Sequencing

Whole genome sequencing can be used to identify germline or somatic mutations, single nucleotide polymorphism (SNPs), indels (insertion and deletion), and copy number variations. In the past 5 years and using NGS, genome-wide association studies have begun to provide unprecedented information of the connection between genetics and disease [45]. For example, such an approach has recently helped to identify new genomic loci for susceptibility to Crohn's disease [46], a chronic debilitating intestinal disease of which the pathogenesis was poorly understood. Not only has the identification of these novel loci improved our understanding of the pathophysiology of the disease but it has also had implications for patient treatment [47]. Genome-wide association studies have also produced data of non-coding sequences implicated in the pathogenesis of complex human disorders [45]. Whole-genome sequencing now permits the compilation of sophisticated databases of full spectrum of germline variants conferring risks for inherited diseases [48] and numerous somatic mutations underlying all aspects of human cancers [47]. The recent $48 million grant from the National Institutes of Health opens the door to the use of NGS to analyze the genomes of thousands of patients who suffer from more than 6,000 rare genetic disorders, many of which follow Mendelian inheritance patterns with mutations involving a single gene (http://www.nih.gov/news/health/dec2011/nhgri-06.htm). It is important to note that as human genomic research has been progressing into the whole-genome sequencing era using NGS, it is imperative to identify and document the genetic variation across human populations to ensure diverse ancestry to be included in our genomic studies so that healthcare disparities introduced by genetics community can be curtailed [49]. Beyond human studies, NGS has been used to study genetic diversities and population structures of endangered animal species [50] and has been found to have significant applications in plant as well [51].

6.2 Targeted Sequencing

NGS has already enabled many forms of clinical diagnostic testing by targeting selected genes or gene exons to accommodate specific clinical needs. Many human disorders are caused by dysfunctions in one of several causative genes. Mutations of genes involved in the same metabolic or signaling pathways may lead to similar disease phenotypes. On the other hand, different mutations involving the same gene may carry subtle to drastically different clinical manifestation of the disease and many diseases may have overlapping mutation profiles. For example, hereditary erythrocyte disorders may involve any of the 27 genes related to the red cell membrane structure, red cell enzyme deficiency, and hemoglobin metabolism [52]. Phenotypic overlap among many involved genes requires precise diagnosis of these disorders by identifying the corresponding gene mutation(s). Indeed, many medical centers have begun to offer clinical mutation analysis using NGS; examples include extensive panels for detection of mutations in one of the 10–30 genes for the diagnosis of cardiomyopathy [53], X-linked congenital diseases [54–58], comprehensive mutation detection in 24 genes known to cause congenital disorders of glycosylation [59], and various other autosomal disorders [60–62].

Whole exome capture and sequencing by NGS have been successfully applied to identify mutations using various tissue sources [63, 64]. Given the magnitude of carrier burden in human population and increasingly available low-cost NGS platforms, targeted carrier screening is also possible in clinical practice to reduce the incidence and suffering in severe recessive childhood disorders [65]. Targeted gene mutation panel analysis for oncology is becoming increasingly important and cost-effective for cancer diagnosis, prognosis, and precision therapy. Currently dozens of major medical centers in the United States are validating such cancer sequencing panels by NGS.

6.3 Epigenetic Applications

Epigenetic applications of NGS may include platforms such as CHIP-seq-Protein-DNA binding and histone modification [66]. Such technology has been used to map the methylome of the diploid human genome recently [67]. Epigenetic applications of NGS are beginning to provide fundamental insights into human biology and diseases, for example the discovery of widespread allele-specific epigenetic variation in the human genome will likely contribute to our understanding of some common diseases with complex genetic background [68, 69].

6.4 Transcriptome Analysis

Through combination of hybridization capture of cDNAs and next-generation sequencing, targeted RNA-Sequencing (RNA-Seq) provides an efficient and cost-effective method to analyze specific subsets of transcriptome simultaneously for mutation, structural alteration, and expression [70]. NGS technologies with appropriate assembly algorithms have facilitated the reconstruction of the entire transcriptome in the absence of a reference genome [71]. Targeted RNA-Seq is also a powerful tool suitable for a wide range of large-scale tumor-profiling studies to identify sequence variations and novel fusion gene products [72].

6.5 Microbial Population Analysis

NGS is ideal for whole viral, bacterial, and yeast genome sequencing owing to its high throughput, depth of sequencing, and appropriate size of most microbial genomes [39]. Currently large amounts of NGS data have become available that will greatly enhance our understanding of host pathogen interactions with the discovery of new transcripts, splice variants, mutations, regulatory elements, and epigenetic controls [73]. For example, NGS was successfully applied to characterize the genome of the German Enterohemorrhagic Escherichial coli O104:H4 outbreak very recently using the Ion Torrent platform [74]. The ability to perform whole-genome comparisons further allows one to link phenotypic dissimilarities among closely related organisms and their underlying genetic mechanisms, and therefore gain a better understanding of pathogen evolution [75]. Current applications of high-throughput whole viral genome sequencing include detection of viral genome variability and evolution within the host (e.g., human immunodeficiency virus and human hepatitis C virus) and monitoring of low-abundance antiviral drug-resistance mutations. NGS techniques lead to a new field of study called "metagenomics." It is now possible to detect unexpected disease-associated viruses and emerging new human viruses, including cancer-related ones [76]. Other applications include HPV typing because of its high detection sensitivity and its broad spectrum coverage of HPV types, subtypes, and variants [77].

7 Limitations of NGS in Clinical Practice

Several technical limitations of NGS in genomic studies or genomic medicine have already been briefed in the aforementioned sections. To emphasize the clinical aspects of NGS application in medicine the following are important confounding factors that are likely the subjects of many future discussions: (1) quality control/ quality assurance programs are likely difficult to be standardized from the initial technical operation to clinical validation; (2) data management and storage (analyzing, managing and instrumentation for storage) require electronic devices of extremely high capacity; (3) daunting challenges in the analysis of sequence data for clinical interpretation (such as previously unknown genetic variants) are major issues to be tackled; (4) reporting complex results may be extremely difficult with regard to clinical implication in disease diagnosis, prognosis and guiding precision therapy; (5) incidental findings with significant biomedical implications may pose ethical responsibilities for pathologists (duty to report); (6) patent infringement may affect laboratories using NGS and reporting genes or DNA sequences under patent protection; and (7) finally how NGS can be adequately reimbursed requires academic institutions, commercial laboratories and regulatory agencies to develop consensus utilities, and fee codes (e.g., CPT codes) in collaboration with and being acceptable by clinicians, service providers and insurance industry.

8 Summary

Defying the Moore's law in computer industry, NGS has been far outperforming its prediction of doubling technical improvement and affordability every 2 years. The exponential improvement of speed and the concomitant astronomical drop in costs are quickly driving NGS from the research arena to the patient bedside. NGS will affect essentially all aspects of clinical care issues enabling many diagnostic tests that have never been considered possible before. In the next 10 years we will likely see the arrival of NGS platforms that are versatile, accurate, affordable, and portable for clinical use. However, a significant hurdle for clinical application of NGS is bioinformatics analysis of the sequencing data. Sequence variant annotation requires data mining into various databases (e.g., locus specific database, HGMD/Biobase, OMIM, SeattleSeq, and 1000 Genome program) and functional prediction programs such as PolyPhen and SIFT are essential for biological and clinical interpretation of new or uncommon sequence variants. Clinical validation of sequence variants in relation to disease or phenotype is more difficult and requires prospective studies of large cohorts of patients. While we transition from single gene analysis in the recent past, to multi-gene panel analysis, to whole exome sequencing, and soon to the whole genome approach, the complexity of the technology and bioinformatics increase dramatically and their clinical applications have been proven far more complicated than previously thought. Navigation from DNA base pairs of the human genome to the bedside of patients will continue to rely on new technologies such as NGS, genomic bioinformatics sciences, large scale collaborative efforts, and a multidisciplinary team approach involving academic institutions, hospitals, government agencies and industries [6].

Acknowledgement and Declaration The author does not have any funding support or financial interest to declare.

References

1. International Human Genome Sequencing Consortium (2004) Finishing the euchromatic sequence of the human genome. Nature 431(7011):931–945
2. Cancer Genome Atlas Research Network (2008) Comprehensive genomic characterization defines human glioblastoma genes and core pathways. Nature 455(7216):1061–1068
3. Ley TJ, Mardis ER, Ding L, Fulton B, McLellan MD, Chen K, Dooling D, Dunford-Shore BH, McGrath S, Hickenbotham M et al (2008) DNA sequencing of a cytogenetically normal acute myeloid leukaemia genome. Nature 456(7218):66–72
4. Pao W, Miller V, Zakowski M, Doherty J, Politi K, Sarkaria I, Singh B, Heelan R, Rusch V, Fulton L et al (2004) EGF receptor gene mutations are common in lung cancers from "never smokers" and are associated with sensitivity of tumors to gefitinib and erlotinib. Proc Natl Acad Sci U S A 101(36):13306–13311
5. Dietz HC (2010) New therapeutic approaches to mendelian disorders. N Engl J Med 363(9):852–863

6. Green ED, Guyer MS (2011) Charting a course for genomic medicine from base pairs to bedside. Nature 470(7333):204–213
7. Hawkins RD, Hon GC, Ren B (2010) Next-generation genomics: an integrative approach. Nat Rev Genet 11(7):476–486
8. Shendure J, Ji H (2008) Next-generation DNA sequencing. Nat Biotechnol 26(10):1135–1145
9. Martinez DA, Nelson MA (2010) The next generation becomes the now generation. PLoS Genet 6(4):e1000906
10. Voelkerding KV, Dames SA, Durtschi JD (2009) Next-generation sequencing: from basic research to diagnostics. Clin Chem 55(4):641–658
11. Wheeler DA, Srinivasan M, Egholm M, Shen Y, Chen L, McGuire A, He W, Chen YJ, Makhijani V, Roth GT et al (2008) The complete genome of an individual by massively parallel DNA sequencing. Nature 452(7189):872–876
12. Niedringhaus TP, Milanova D, Kerby MB, Snyder MP, Barron AE (2011) Landscape of next-generation sequencing technologies. Anal Chem 83(12):4327–4341
13. Fuller CW, Middendorf LR, Benner SA, Church GM, Harris T, Huang X, Jovanovich SB, Nelson JR, Schloss JA, Schwartz DC et al (2009) The challenges of sequencing by synthesis. Nat Biotechnol 27(11):1013–1023
14. Ekblom R, Galindo J (2011) Applications of next generation sequencing in molecular ecology of non-model organisms. Heredity 107(1):1–15
15. Sanger F, Nicklen S, Coulson AR (1977) DNA sequencing with chain-terminating inhibitors. Proc Natl Acad Sci U S A 74(12):5463–5467
16. Maxam AM, Gilbert W (1977) A new method for sequencing DNA. Proc Natl Acad Sci U S A 74(2):560–564
17. Levy S, Sutton G, Ng PC, Feuk L, Halpern AL, Walenz BP, Axelrod N, Huang J, Kirkness EF, Denisov G et al (2007) The diploid genome sequence of an individual human. PLoS Biol 5(10):e254
18. Metzker ML (2010) Sequencing technologies – the next generation. Nat Rev Genet 11 (1):31–46
19. Margulies M, Egholm M, Altman WE, Attiya S, Bader JS, Bemben LA, Berka J, Braverman MS, Chen YJ, Chen Z et al (2005) Genome sequencing in microfabricated high-density picolitre reactors. Nature 437(7057):376–380
20. Schuster SC (2008) Next-generation sequencing transforms today's biology. Nat Methods 5(1):16–18
21. Ronaghi M, Uhlen M, Nyren P (1998) A sequencing method based on real-time pyrophosphate. Science 281(5375):363–365
22. Mardis ER (2008) Next-generation DNA sequencing methods. Annu Rev Genomics Hum Genet 9:387–402
23. Valouev A, Ichikawa J, Tonthat T, Stuart J, Ranade S, Peckham H, Zeng K, Malek JA, Costa G, McKernan K et al (2008) A high-resolution, nucleosome position map of C. elegans reveals a lack of universal sequence-dictated positioning. Genome Res 18(7):1051–1063
24. Wu H, Irizarry RA, Bravo HC (2010) Intensity normalization improves color calling in SOLiD sequencing. Nat Methods 7(5):336–337
25. Efcavitch JW, Thompson JF (2010) Single-molecule DNA analysis. Annu Rev Anal Chem (Palo Alto Calif) 3:109–128
26. Thompson JF, Milos PM (2011) The properties and applications of single-molecule DNA sequencing. Genome Biol 12(2):217
27. Harris TD, Buzby PR, Babcock H, Beer E, Bowers J, Braslavsky I, Causey M, Colonell J, Dimeo J, Efcavitch JW et al (2008) Single-molecule DNA sequencing of a viral genome. Science 320(5872):106–109
28. Eid J, Fehr A, Gray J, Luong K, Lyle J, Otto G, Peluso P, Rank D, Baybayan P, Bettman B et al (2009) Real-time DNA sequencing from single polymerase molecules. Science 323 (5910):133–138

29. Lundquist PM, Zhong CF, Zhao P, Tomaney AB, Peluso PS, Dixon J, Bettman B, Lacroix Y, Kwo DP, McCullough E et al (2008) Parallel confocal detection of single molecules in real time. Opt Lett 33(9):1026–1028
30. Korlach J, Marks PJ, Cicero RL, Gray JJ, Murphy DL, Roitman DB, Pham TT, Otto GA, Foquet M, Turner SW (2008) Selective aluminum passivation for targeted immobilization of single DNA polymerase molecules in zero-mode waveguide nanostructures. Proc Natl Acad Sci U S A 105(4):1176–1181
31. Korlach J, Bjornson KP, Chaudhuri BP, Cicero RL, Flusberg BA, Gray JJ, Holden D, Saxena R, Wegener J, Turner SW (2010) Real-time DNA sequencing from single polymerase molecules. Methods Enzymol 472:431–455
32. Munroe DJ, Harris TJ (2010) Third-generation sequencing fireworks at Marco Island. Nat Biotechnol 28(5):426–428
33. Flusberg BA, Webster DR, Lee JH, Travers KJ, Olivares EC, Clark TA, Korlach J, Turner SW (2010) Direct detection of DNA methylation during single-molecule, real-time sequencing. Nat Methods 7(6):461–465
34. Uemura S, Aitken CE, Korlach J, Flusberg BA, Turner SW, Puglisi JD (2010) Real-time tRNA transit on single translating ribosomes at codon resolution. Nature 464(7291):1012–1017
35. Drmanac R, Sparks AB, Callow MJ, Halpern AL, Burns NL, Kermani BG, Carnevali P, Nazarenko I, Nilsen GB, Yeung G et al (2010) Human genome sequencing using unchained base reads on self-assembling DNA nanoarrays. Science 327(5961):78–81
36. Porreca GJ (2010) Genome sequencing on nanoballs. Nat Biotechnol 28(1):43–44
37. Roach JC, Glusman G, Smit AF, Huff CD, Hubley R, Shannon PT, Rowen L, Pant KP, Goodman N, Bamshad M et al (2010) Analysis of genetic inheritance in a family quartet by whole-genome sequencing. Science 328(5978):636–639
38. Lee W, Jiang Z, Liu J, Haverty PM, Guan Y, Stinson J, Yue P, Zhang Y, Pant KP, Bhatt D et al (2010) The mutation spectrum revealed by paired genome sequences from a lung cancer patient. Nature 465(7297):473–477
39. Rothberg JM, Hinz W, Rearick TM, Schultz J, Mileski W, Davey M, Leamon JH, Johnson K, Milgrew MJ, Edwards M et al (2011) An integrated semiconductor device enabling non-optical genome sequencing. Nature 475(7356):348–352
40. Clarke J, Wu HC, Jayasinghe L, Patel A, Reid S, Bayley H (2009) Continuous base identification for single-molecule nanopore DNA sequencing. Nat Nanotechnol 4(4):265–270
41. Stoddart D, Heron AJ, Mikhailova E, Maglia G, Bayley H (2009) Single-nucleotide discrimination in immobilized DNA oligonucleotides with a biological nanopore. Proc Natl Acad Sci U S A 106(19):7702–7707
42. Branton D, Deamer DW, Marziali A, Bayley H, Benner SA, Butler T, Di Ventra M, Garaj S, Hibbs A, Huang X et al (2008) The potential and challenges of nanopore sequencing. Nat Biotechnol 26(10):1146–1153
43. Bayley H (2006) Sequencing single molecules of DNA. Curr Opin Chem Biol 10(6):628–637
44. Astier Y, Braha O, Bayley H (2006) Toward single molecule DNA sequencing: direct identification of ribonucleoside and deoxyribonucleoside 5'-monophosphates by using an engineered protein nanopore equipped with a molecular adapter. J Am Chem Soc 128(5):1705–1710
45. Manolio TA (2010) Genomewide association studies and assessment of the risk of disease. N Engl J Med 363(2):166–176
46. Rioux JD, Xavier RJ, Taylor KD, Silverberg MS, Goyette P, Huett A, Green T, Kuballa P, Barmada MM, Datta LW et al (2007) Genome-wide association study identifies new susceptibility loci for Crohn disease and implicates autophagy in disease pathogenesis. Nat Genet 39(5):596–604
47. Van Limbergen J, Wilson DC, Satsangi J (2009) The genetics of Crohn's disease. Annu Rev Genomics Hum Genet 10:89–116
48. Hoffmann TJ, Kvale MN, Hesselson SE, Zhan Y, Aquino C, Cao Y, Cawley S, Chung E, Connell S, Eshragh J et al (2011) Next generation genome-wide association tool: design and coverage of a high-throughput European-optimized SNP array. Genomics 98(2):79–89

49. Need AC, Goldstein DB (2009) Next generation disparities in human genomics: concerns and remedies. Trends Genet 25(11):489–494
50. Miller W, Hayes VM, Ratan A, Petersen DC, Wittekindt NE, Miller J, Walenz B, Knight J, Qi J, Zhao F et al (2011) Genetic diversity and population structure of the endangered marsupial Sarcophilus harrisii (Tasmanian devil). Proc Natl Acad Sci U S A 108(30):12348–12353
51. Zalapa JE, Cuevas H, Zhu H, Steffan S, Senalik D, Zeldin E, McCown B, Harbut R, Simon P (2012) Using next-generation sequencing approaches to isolate simple sequence repeat (SSR) loci in the plant sciences. Am J Bot 99:193–208
52. Mohandas N, Gallagher PG (2008) Red cell membrane: past, present, and future. Blood 112 (10):3939–3948
53. Hershberger RE, Siegfried JD (2011) Update 2011: clinical and genetic issues in familial dilated cardiomyopathy. J Am Coll Cardiol 57(16):1641–1649
54. Kingsmore SF, Dinwiddie DL, Miller NA, Soden SE, Saunders CJ (2011) Adopting orphans: comprehensive genetic testing of Mendelian diseases of childhood by next-generation sequencing. Expert Rev Mol Diagn 11(8):855–868
55. Tsurusaki Y, Okamoto N, Suzuki Y, Doi H, Saitsu H, Miyake N, Matsumoto N (2011) Exome sequencing of two patients in a family with atypical X-linked leukodystrophy. Clin Genet 80(2):161–166
56. Tsurusaki Y, Osaka H, Hamanoue H, Shimbo H, Tsuji M, Doi H, Saitsu H, Matsumoto N, Miyake N (2011) Rapid detection of a mutation causing X-linked leucoencephalopathy by exome sequencing. J Med Genet 48(9):606–609
57. Schraders M, Haas SA, Weegerink NJ, Oostrik J, Hu H, Hoefsloot LH, Kannan S, Huygen PL, Pennings RJ, Admiraal RJ et al (2011) Next-generation sequencing identifies mutations of SMPX, which encodes the small muscle protein, X-linked, as a cause of progressive hearing impairment. Am J Hum Genet 88(5):628–634
58. Hu H, Wrogemann K, Kalscheuer V, Tzschach A, Richard H, Haas SA, Menzel C, Bienek M, Froyen G, Raynaud M et al (2009) Mutation screening in 86 known X-linked mental retardation genes by droplet-based multiplex PCR and massive parallel sequencing. Hugo J 3 (1–4):41–49
59. Jones MA, Bhide S, Chin E, Ng BG, Rhodenizer D, Zhang VW, Sun JJ, Tanner A, Freeze HH, Hegde MR (2011) Targeted polymerase chain reaction-based enrichment and next generation sequencing for diagnostic testing of congenital disorders of glycosylation. Genet Med 13 (11):921–932
60. Doi H, Yoshida K, Yasuda T, Fukuda M, Fukuda Y, Morita H, Ikeda S, Kato R, Tsurusaki Y, Miyake N et al (2011) Exome sequencing reveals a homozygous SYT14 mutation in adult-onset, autosomal-recessive spinocerebellar ataxia with psychomotor retardation. Am J Hum Genet 89(2):320–327
61. Artuso R, Fallerini C, Dosa L, Scionti F, Clementi M, Garosi G, Massella L, Epistolato MC, Mancini R, Mari F et al (2012) Advances in Alport syndrome diagnosis using next-generation sequencing. Eur J Hum Genet 20(1):50–57
62. Bowne SJ, Sullivan LS, Koboldt DC, Ding L, Fulton R, Abbott RM, Sodergren EJ, Birch DG, Wheaton DH, Heckenlively JR et al (2011) Identification of disease-causing mutations in autosomal dominant retinitis pigmentosa (adRP) using next-generation DNA sequencing. Invest Ophthalmol Vis Sci 52(1):494–503
63. Choi M, Scholl UI, Ji W, Liu T, Tikhonova IR, Zumbo P, Nayir A, Bakkaloglu A, Ozen S, Sanjad S et al (2009) Genetic diagnosis by whole exome capture and massively parallel DNA sequencing. Proc Natl Acad Sci U S A 106(45):19096–19101
64. Choi M, Scholl UI, Yue P, Bjorklund P, Zhao B, Nelson-Williams C, Ji W, Cho Y, Patel A, Men CJ et al (2011) K+ channel mutations in adrenal aldosterone-producing adenomas and hereditary hypertension. Science 331(6018):768–772
65. Bell CJ, Dinwiddie DL, Miller NA, Hateley SL, Ganusova EE, Mudge J, Langley RJ, Zhang L, Lee CC, Schilkey FD et al (2011) Carrier testing for severe childhood recessive diseases by next-generation sequencing. Sci Transl Med 3(65):65ra4

66. Park PJ (2009) ChIP-seq: advantages and challenges of a maturing technology. Nat Rev Genet 10(10):669–680
67. Fouse SD, Nagarajan RP, Costello JF (2010) Genome-scale DNA methylation analysis. Epigenomics 2(1):105–117
68. Meaburn E, Schulz R (2011) Next generation sequencing in epigenetics: insights and challenges. Semin Cell Dev Biol 23:192–199
69. Ku CS, Naidoo N, Wu M, Soong R (2011) Studying the epigenome using next generation sequencing. J Med Genet 48(11):721–730
70. Gibbons JG, Janson EM, Hittinger CT, Johnston M, Abbot P, Rokas A (2009) Benchmarking next-generation transcriptome sequencing for functional and evolutionary genomics. Mol Biol Evol 26(12):2731–2744
71. Martin JA, Wang Z (2011) Next-generation transcriptome assembly. Nat Rev Genet 12 (10):671–682
72. Levin JZ, Berger MF, Adiconis X, Rogov P, Melnikov A, Fennell T, Nusbaum C, Garraway LA, Gnirke A (2009) Targeted next-generation sequencing of a cancer transcriptome enhances detection of sequence variants and novel fusion transcripts. Genome Biol 10(10):R115
73. Tripathy S, Jiang RH (2012) Massively parallel sequencing technology in pathogenic microbes. Methods Mol Biol 835:271–294
74. Mellmann A, Harmsen D, Cummings CA, Zentz EB, Leopold SR, Rico A, Prior K, Szczepanowski R, Ji Y, Zhang W et al (2011) Prospective genomic characterization of the German enterohemorrhagic Escherichia coli O104:H4 outbreak by rapid next generation sequencing technology. PLoS One 6(7):e22751
75. Hu B, Xie G, Lo CC, Starkenburg SR, Chain PS (2011) Pathogen comparative genomics in the next-generation sequencing era: genome alignments, pangenomics and metagenomics. Brief Funct Genomics 10(6):322–333
76. Barzon L, Lavezzo E, Militello V, Toppo S, Palu G (2011) Applications of next-generation sequencing technologies to diagnostic virology. Int J Mol Sci 12(11):7861–7884
77. Barzon L, Militello V, Lavezzo E, Franchin E, Peta E, Squarzon L, Trevisan M, Pagni S, Dal Bello F, Toppo S et al (2011) Human papillomavirus genotyping by 454 next generation sequencing technology. J Clin Virol 52(2):93–97

Top Curr Chem (2014) 336: 19–46
DOI: 10.1007/128_2012_325

Published online: 11 May 2012

Application of Next Generation Sequencing to Molecular Diagnosis of Inherited Diseases

Wei Zhang, Hong Cui, and Lee-Jun C. Wong

Abstract Recent development of high throughput, massively parallel sequencing (MPS or next generation sequencing, NGS) technology has revolutionized the molecular diagnosis of human genetic disease. The ability to generate enormous amount of sequence data in a short time at an affordable cost makes this approach ideal for a wide range of applications from sequencing a group of candidate genes, all coding regions (known as exome sequencing) to the entire human genome. The technology brings about an unprecedented application to the identification of the molecular basis of hard-to-diagnose genetic disorders. This chapter reviews the up-to-date published application of next generation sequencing in clinical molecular diagnostic laboratories. We also emphasize the various target gene enrichment methods and their advantages and shortcomings. Obstacles to compliance with regulatory authorities like CLIA/CAP in clinical settings are also briefly discussed.

Keywords Human genetic diseases, Massively parallel sequencing (MPS), Molecular diagnosis, Next generation sequencing (NGS), Target gene enrichment

Contents

1 Introduction 20
2 Current Clinical Application of MPS 21
  2.1 Target Gene(s) Sequencing in the Clinical Setting 21
  2.2 Whole Exome Sequencing 31
  2.3 Whole Genome Sequencing 32
  2.4 Application of MPS to Prenatal Diagnosis 33

W. Zhang, H. Cui and L.-J.C. Wong (✉)
Department of Molecular and Human Genetics, Baylor College of Medicine, One Baylor Plaza, NAB 2015, Houston TX 77030, USA
e-mail: ljwong@bcm.edu

3 Target Gene Enrichment Methods 33
3.1 Multiplex PCR 33
3.2 Oligonucleotide Probe Based Capture on a Solid Phase 34
3.3 Oligonucleotide Probe Based Capture in Solution Phase 35
4 Cautionary Tale of Current MPS Tests Offered for Clinical Application 35
4.1 PCR-Based Target Amplification Has Intrinsic Drawbacks 35
4.2 Lack of Sufficient Clinical Validation, Specificity, Sensitivity 36
4.3 High False Negative Rate Due to Low Coverage 36
4.4 Confirmation of Detected Variants to Rule Out False Positives 37
4.5 Interpretation of Novel Variants: The Most Challenging Task 37
5 Discussion 38
6 Conclusion 39
References 39

1 Introduction

Over the past 20–30 years, DNA sequencing by the Sanger method has been regarded as the gold standard for the identification of aberrant DNA sequence changes (mutations, in genetic terms) to support a genetic disease diagnosis [1]. For single gene disorders with clear clinical/biochemical indication and/or known mutation hot spots, Sanger sequencing of the target region is an accurate and cost effective way to obtain a definitive molecular diagnosis. However, the selection of candidate gene(s) for sequence analysis is extremely difficult, as most inherited disorders exhibit genetic and clinical heterogeneity. Thus, it often takes a long time to obtain the right molecular diagnosis by stepwise sequencing the probable causative genes with consequent high cost and anxiety for the patient's family. An example of an extreme situation is the diagnosis of mitochondrial disorders, for which there are significant clinical overlaps and over 1,300 genes are responsible [2–5]. In recent years, owing to the fast development of a variety of sequencing technologies in the post human genome project era, large scale sequencing such as (1) a group of target genes, (2) all of protein coding regions of the human genome, and (3) even the whole human genome have become a reality [6–12]. Next Generation Sequencing (NGS) or Massively Parallel Sequencing (MPS) offers a way to detect mutations in many different genes in a cost and time efficient manner through their ability to generate deep coverage of the target sequences [13–19].

Various sequencing platforms using different sequencing chemistry have been developed [20]. The three major currently commercially available platforms that produce gigabases of output are Illumina's sequencing by DNA synthesis, Roche's 454 by pyro-sequencing, and ABI SOLiD by oligonucleotide ligation [21–24]. There are also platforms for smaller scale sequencers, such as Ion Torrent, which is based on semiconductors to detect protons generated by polymerase reactions [25], single molecule sequencing by Helicos Helioscope [26], and Pacific Bioscience's single molecular real time (SMRT) instrument [27, 28].

Regardless of the chemistry and platforms, these methods share one common principle, which is to sequence numerous spatially separated genomic regions in a massively parallel manner [13, 20, 29]. The detailed chemistry, sequencing principles, their hardware, and their advantages and pitfalls are beyond the scope of this chapter but are discussed in chapter 1 of this book.

The MPS approaches have been successfully applied to many different research studies to identify new disease genes [30], mutations in non-coding regions [31–34], and epigenetic changes in the whole genome [35, 36]. Similar approaches have been applied to molecular diagnosis of inherited disorders, particularly complex diseases with heterogeneous clinical phenotype and multiple underlying genetic causes [37–40]. However, it is still not economically feasible and is technically demanding to sequence the 3×10^9 base pairs of the whole human genome. Therefore, in the context of clinical applications, it is often desirable to capture or enrich a group of genes known to be responsible for a certain type of clinical phenotypes, followed by MPS. Good examples are genes causing cardiomyopathy or mitochondrial respiratory chain disorders [38, 40]. Therefore, depending on the amount of sequence to be analyzed, the purpose of the analysis, and the available sequencing platforms, the method of target gene enrichment may vary greatly. In addition, in order to use newly developed MPS tests for clinical diagnosis, stringent validation and quality control procedures need to be instituted in compliance with regulating authorities, such as CAP or CLIA to assure quality of service [41–43]. It is likely that all laboratory-developed tests (LDTs) will be subjected to regulation in the near future in many developed countries [44].

In this chapter we will review the reported clinical applications of MPS to the molecular diagnosis of inherited disorders, various gene enrichment methods, potential pitfalls, necessary quality control procedures, and remaining obstacles for comprehensive genomic analysis using MPS.

2 Current Clinical Application of MPS

A review of the recent publications on the introduction of MPS technology into clinical practice has revealed the broad utility of this novel technology in the molecular diagnosis of a variety of human genetic disorders (Table 1). The clinical applications vary from single gene disorders such as neurofibromatosis Type 1 (NF1), Marfan syndrome (MFS), and spastic paraplegia [45–47] to diseases caused by a group of related genes such as hypertrophic cardiomyopathy and congenital disorders of glycosylation (CDG) [38, 39, 48]. MPS has also been applied to multi-gene disorders including X-linked intellectual disability (XLID) and retinitis pigmentosa [37, 49] as well as defined disorders without identified genetic causes [9, 52, 53]. Various gene enrichment methods were employed (Table 2) depending on the purpose of application.

2.1 Target Gene(s) Sequencing in the Clinical Setting

MPS has been applied to assist with molecular diagnosis of well-defined disorders caused by a group of genes. In order to sequence specifically the regions of interest, these target genes need to be enriched tens of thousands of times or more to avoid the interference of the remaining bulk of the genome. Depending on the amount of target sequences, the enrichment method may vary.

Table 1 Summary of the recent publications on the introduction of MPS technology to clinical practice

	Gene/disease	Enrichment method	MPS Sequencing platform/ chemistry	Sample	Average coverage	% reads mapped to target	Analysis software	References
1	*NF1*	NimbleGen oligo array capture	GS-FLX (SBS, pyro-sequencing)	2 known	>30X	~52% 59% of 52% mapped to ch 17 *NF1*	NextGENe	[45]
2	Marfan+LDS *FBN1* +*TGFBR1*+ *TGFBR2*	Multiplex PCR	GS-FLX (SBS, pyro-sequencing)	5 known 87 unknown	~174X	55–85%	In house variant interpret pipeline	[46]
3	*CYP7B1* (SPG5) +*SPG7*	Fluidigm (FD)	GS-FLX (SBS, pyro-sequencing)	187 patients	72X for run1 25X for run2	80% of target exons have more than 20X coverage	SeqNext v3.4.1 Build 504 (JSI German)	[47]
4	DCM 19 genes	Pooled PCR amplicons	GAII (SBS)	5 known	~50X	59–69%	NextGENe	[38]
5	CDG 24 gene	Fluidigm (FD) RainDance (RD)	SOLiD version 3/50 base read, SE (SBL)	12 known	616X (FD) 455X (RD)	48%	NextGENe	[48]
6	RP 45 genes	NimbleGen oligo array capture	GAII/32 base read, SE (SBS)	2 known 3 unknown	486X (1 sample per lane) 98X (4 samples per lane)	~35%	Genomic Workbench	[49]
7	XLMR 86 genes	RainDance (RD)	GAII (SBS)	3 known 21 unknown	Coverage per base ranging from 92X to 445X	67.9%	SOAP2.20	[37]
8	mtDNA	2 overlapping PCR fragments	GAII (SBS)	2 known	~1,785x	N/A	DNAStar & NextGENe	[50]
9	mtDNA and 362 nuclear genes	Agilent array-based capture	GAII/36 base read, SE (SBS)	2 patients 1 normal	37X-51X for nuclear genes, 3,000–5,000X for mtDNA	17–35% mapped to nuclear genome, 20–37% mapped to mtDNA	MAQ	[40]

10	Carrier test of 437 genes	Agilent solution-based capture RainDance (RD)	GAII/50 base read (SBS) SOLiD3/50 base read (SBL) HiSeq/130 base read, PE (SBS)	104 known	Varies between different platforms	Varies from 13.7% to 31.7%	GSNAP (for SBS data) BioScope v1.2 (for SBL data)	[51]
11	Miller syndrome WES	Agilent array-based capture	GAII/76 base read, SE (SBS)	3 kindreds	40X	97%	MAQ	[52]
12	Kabuki syndrome WES	Agilent array-based capture	GAII/SE or PE (SBS)	10 unknown	40X	N/A	MAQ, Phaster v1.100122a	[53]
13	Inflammatory bowel disease (IBD) WES	NimbleGen exome array-based capture	GS-FLX (SBS, pyro-sequencing)	1 patient	34X	76.2%	gsMapper (Roche 454)	[54]
14	Charcot–Marie–Tooth (CMT) WGS	Direct genomic DNA	SOLiD (SBL)	Family members	30X	N/A	In house variant interpret pipeline	[9]
15	Dopa-responsive dystonia (DRD) WGS	Direct genomic DNA	SOLiD4 (SBL)	Twins and family members	30X	N/A	In house variant interpret pipeline	[11]
16	Prenatal diagnosis	Direct plasma DNA	SOLiD3 (SBL)	15 unknown	N/A	21%	SOLiD alignment tool	[55]

SE single-end, *PE* paired-end, *SBS* sequencing-by-synthesis, *SBL* sequencing-by-ligation, *WES* whole exome sequencing, *WGS* whole genome sequencing

Table 2 Comparison of different gene enrichment methods

Method	Target size	Throughput	Automation adaptability	Advantage	Disadvantage
PCR-based enrichment (commercial platform)					
Regular PCR/ multiplex	Small	Low	Yes	Simple, straight forward Flexible in optimizing conditions	Cost-inefficient Limited target size
Array PCR (Fluidigm)	Small to medium	Low	No	Cost-efficient reactions Easy to perform Uniform PCR condition	High cost in synthesizing primers Varied amplification efficiency Fixed format
Microdroplet PCR (RainDance)	Medium	Low	Yes	Up to 20,000 amplicons Relatively equal amplification efficiency	High cost in synthesizing primers Varied amplification efficiency Need special equipment Less flexibility
Capture-based enrichment					
Oligonucleotide array-based capture	Medium to high	Low	No	Target size is expandable Low cost per gene	Need one extra step to elute the captured DNA Need special equipment Missing GC rich regions
Oligonucleotide Solution-based capture	Medium to high	High	Yes	Target size is expandable Low cost per gene Easy for automation	Cost-inefficient for small target region Missing GC rich regions

2.1.1 Examples of Single Gene Disorders and the Target Enrichment Approach

Neurofibromatosis Type 1 (NF1)

Neurofibromatosis type 1 is an autosomal dominant disorder caused by mutations in the *NF1* (17q11.2) gene. It is one of the most commonly inherited genetic disorders with an estimated prevalence of 1 in 3,000 individuals [56]. Molecular defects in the

NF1 gene include a variety of mutation types from point mutations and small indels to large complex rearrangements [57]. In addition, the presence of pseudogene complicates the sequencing analysis of the target gene. A new molecular analysis was carried out with a microarray based target gene capture and enrichment method, followed by MPS [45]. In this report, DNA samples from two patients with a definitive diagnosis of neurofibromatosis type 1 and known *NF1* mutations were analyzed. The 282 kb *NF1* gene region was fragmented and captured on an array containing specific oligonucleotide probes followed by sequencing using Roche (454) Genome Sequencer FLX system. About 59% of mapped reads were localized to the targeted *NF1* loci on chromosome 17 in both cases with a depth of coverage >30X. After filtering out the false positive calls introduced by low sequencing coverage, and manual removal of pseudogene, an Alu sequence insertion and a frame-shift single base deletion were identified in each of the two patients respectively. However, only 3 out of 17 and 3 out of 9 identified variants in each of these two samples respectively were confirmed by Sanger sequencing [45]. It indicated that sequence variants identified by MPS might have a high percentage of false positives.

Marfan Syndrome

MPS has also been applied to the molecular diagnosis of MFS [46]. MFS is a common (1 in 5,000) autosomal dominant connective tissue disorder, characterized by tall stature with long limbs, dislocated lens, and severe cardiovascular manifestation including aorta aneurysms [58, 59]. The majority of mutations causing MFS are missense, nonsense, and small insertion/deletions in the *FBN1* gene encoding the fibrillin-1 protein. Due to the large number of exons, sequencing by PCR/Sanger is very time consuming and costly. Baetens et al. developed a multiplexed PCR based-MPS approach to sequence simultaneously all exons in 87 samples [46]. In addition to *FBN1*, transforming growth factor beta receptor genes *TGFBR1* and *TGFBR2* are also included. A total of 117 amplicons were amplified in 17 multiplex PCR sets, each containing 4–11 amplicons and sequencing was performed on the Roche (454) Genome Sequencer FLX system. Five samples from known patients diagnosed with MFS or LDS (Loeys–Dietz Syndrome) were selected for a pilot study, which confirmed 23 out of the total 25 calls made by Sanger sequencing, leaving 2 variants unidentified due to low coverage. Additionally, 20 false positive variants were called, all residing in homopolymeric regions. Consequently, to rule out false positives, 13 homopolymer regions were sequenced by the Sanger method separately. Subsequent analysis of 87 patients with typical MFS identified 75 *FBN1* mutations by MPS, equivalent to an 86% (75/87) detection rate.

Hereditary Spastic Paralegias

Hereditary spastic paralegias (HSPs) are a group of inherited neurodegenerative disorders characterized by weakness and spasticity in the lower limbs. They are genetically heterogeneous and at least 46 genetic loci have been identified so far, showing autosomal dominant, autosomal recessive, or X-linked form of inheritance [60]. In the study carried out by Schlipf and coworkers, 2 genes, *CYP7B1* (*SPG5*) and *SPG7*, that are involved in the autosomal recessive form of HSP, were evaluated in 187 patients using MPS [47]. Target gene regions were enriched by an array based microfluidic chip amplification method (Fluidigm) followed by next-generation pyro-sequencing using the Roche (454) Genome Sequencer FLX system. Results from two independent runs showed that 80% of the exons were covered at more than 20X. The remaining poorly covered regions were analyzed by Sanger sequencing separately. Mutations were identified in a total of ten patients: three with *SPG5* mutations and seven with *SPG7* mutations.

2.1.2 Panels of Target Genes for a Disease that Could Be Caused by Mutations in One of Many Potentially Responsible Genes

Dilated Cardiomyopathy

Dilated cardiomyopathy (DCM) is a group of genetically heterogeneous disorders with more than 30 genes identified so far. For three decades, Sanger sequencing of the genes one-by-one has been used, but this is expensive, and time consuming. Resequencing arrays has been considered as an alternative method, but it is well known that they cannot reliably detect insertions and deletions or mutations in high GC rich regions. Additionally, if there were a need for revision in the original design, the cost of making a small change is not economical for most laboratories. Thus, it is impractical to expand quickly the array content as new genes are discovered [61, 62]. These obstacles are likely to be solved by the newly developed MPS technologies that allow simultaneous sequencing of a large number of genes. Gowrisankar and coworkers applied this method to analyze a subset of 19 genes (*ABCC9, ACTC, ACTN2, CSRP3, CTF1, DES, EMD, LDB3, LMNA, MYBPC3, MYH7, PLN, SGCD, TAZ, TCAP, TNNI3, TNNT2, TPM1*, and *VCL*) known to cause DCM. The target coding exons were amplified by the PCR method and their products were pooled for library construction [38]. Five samples with known mutations were validated by this method on the Illumina Genome Analyzer II (GAII), a first generation MPS platform. Results were compared to those of Sanger sequencing and an array-based resequencing chip [38].

An estimated false positive rate of 10.8% and false negative rate of 3.4% were observed. Exons with poor coverage were clearly noticeable, which was the primary reason for missing calls (false negative). The importance of confirmation by a second method to remove the false positive variant calls was emphasized. About 9.3% (23 out of 246) of poorly covered amplicons required Sanger sequencing to minimize the percentage of missed calls. The authors concluded that even though the quality of MPS based testing for the set of 19 genes related to DCM is better compared to the array based re-sequencing method, but the cost associated with test setup and the turnaround time need to be improved before implementing as a routine clinical testing [38]. At the time of writing this review, clinical MPS DCM testing has been already available from a few clinical laboratories.

Congenital Disorders of Glycosylation

The congenital disorders of glycosylation (CDG) are a group of diseases caused by molecular defects in more than 30 genes involved in the N-linked glycosylation pathway [63]. New underlying genetic causes for CDG are continuously being discovered [64, 65]. The prevalence of CDG is estimated to be 1 in 20,000. The disease is usually devastating [63, 66]. Approximately 40% of CDG patients do not have a definitive molecular diagnosis, which is required for the prenatal diagnosis for the affected families. A newly developed MPS-based method was designed to analyze a panel of 24 genes (*ALG2, ALG3, ALG6, ALG8, ALG9, ALG12, ATP6V0A2, B4GALT1, COG1, COG7, COG8, DOLK, DPAGT1, DPM1, GNE, MGAT2, MOGS, MPDUI, MPI, PMM2, RFT1, SLC35A1, SLC35C1*, and *TUSC3)* known to cause CDGs [48].

A total of 215 coding exons were enriched by two different PCR methods: the microdroplet-based PCR (RainDance, RD) with custom designed primers and multiplex PCR by the microfluidic chip (Fluidigm, FD) with in-house designed primers. PCR products were pooled and subjected to ABI SOLiD library preparation and sequencing by ligation [24].

Twelve samples from patients with a known diagnosis of CDG were sequenced for validation using the methods described above. The results showed that about 26–32% of the total variant calls appear to depend on the target enrichment methods. A total of 455 and 616 variants were called by RD and FD methods, respectively. After filtering the data to eliminate the low coverage and low quality calls, 85% and 94% of the unique variants called by the RD and FD, respectively, were likely false positives. In contrast, only 27% of the variants called by both methods were likely not real. About 28 exons (ca. >10% library failure rate) had low or no coverage, which required Sanger sequencing to fill these gaps. The false-negative rate could not be determined due to the lack of DNA [48].

This 24 CDG gene analysis demonstrated the cost-effectiveness of highly multiplexed analysis vs the conventional step-wise single gene approach. The authors also emphasized the importance of verification of all clinically significant variants by Sanger sequencing [48].

Retinitis Pigmentosa

Similar to DCM and CDG, retinitis pigmentosa (RP) is also a group of genetically diverse diseases caused by mutations in any one of more than 40 known genes. Diagnosis of RP is complicated by the lack of clear clinical indication for a specific genetic defect. This disease can be inherited as autosomal dominant, autosomal recessive, or X-linked. About 50% of all patients with RP are sporadic simplex cases and it is clinically difficult to pinpoint the causal genes for molecular diagnosis [67]. The recent development of MPS overcomes these obstacles.

Oligonucleotide DNA probes for a total of 681 coding exons encompassing 359 kb regions of a panel of 45 RP genes were printed onto a custom-designed solid-phase capture array [49]. Target sequences of DNA samples from five patients with RP were enriched by array capture followed by MPS using Genome Analyzer II with 40 bp single end reads [49].

This study identified one patient with compound heterozygous missense mutations in the *CRB1* gene, a second patient with a known homozygous missense mutation in the *PDE6B* gene, and a third patient with a homozygous novel missense variant in the *CNGB1* gene that is predicted to be deleterious. Although predicted deleterious variants were found in the remaining two patients, due to non-segregation of these variants with the disease the molecular diagnosis could not be confirmed.

In this study, if one sample was analyzed in one lane of the flow cell, the mean coverage was 486X with 99% regions covered at >20X, compared to the mean coverage of 98X and 88% regions with >20X coverage by pooling four samples. A total of 582 variants were detected, 150 of them novel. Only the possibly pathogenic variants were verified by Sanger sequencing. The data were not sufficient to establish false positive and false negative rates. The authors emphasized the importance of achieving the requirement of >15X coverage in at least 90% of target regions as a clinical standard and also suggested outsourcing as an alternative to lower the cost [49].

X-Linked Intellectual Disability

As much as 2–3% of the general population meets clinical criteria of intellectual disability. XLID is very heterogeneous. More than 100 XLID genes have been identified, which account for only a portion of families with unambiguous XLID [68]. Using MPS, analysis of a panel of 86 XLID genes has been offered as a clinical molecular test in order to improve the detection rate.

The target regions of the 86 XLID genes of 24 unrelated patients, each from a large XLID family, were enriched by microdroplet-based multiplex PCR (RainDance) with custom designed primer sets for 1,912 amplicons; MPS was performed on Illumina GA II [37].

This amplicon-based approach revealed that 88.5% of targeted regions were covered with a median coverage of 249X. Between 87.9% and 99.5% of targeted regions were covered by sequence reads with a base quality score ≥30 at a minimal coverage of 2X [37]. It was reported that 131 out of 1,912 amplicons cannot be unambiguously mapped, and 34 amplicons failed to amplify, corresponding to an 8.6% failure rate.

The most noticeable is the large variations (from 12.2% to 57.4%) in the percentage of reads mapped to the targeted regions. The authors attributed the variation to different amplification efficiencies in different samples. In addition to three known positive controls in the 24 samples, deleterious mutations were identified in 7 additional patients, with a detection rate of 33% (7/21). Nevertheless, the authors considered this method to be robust and usable as a screening test for XLID [37].

2.1.3 Sequencing the Mitochondrial Genome and Autosomal Genes Coding for Mitochondrial Proteins

In addition to 23 pairs of nuclear chromosomes, each human cell contains hundreds to thousands of copies of the mitochondrial genome. The human mitochondrial genome is a 16.6-kb circular double stranded DNA encoding 37 genes: 22 tRNA, 2 rRNA, and 13 mRNAs for 13 respiratory chain complex protein subunits. Mutations in mitochondrial DNA (mtDNA) cause a broad clinical spectrum of respiratory chain disorders with the tissues of high energy demand such as brain and muscle preferentially affected [69]. Disease severity depends on the degree of mutation heteroplasmy in the affected tissues [69]. Therefore, diagnosis of mtDNA disorders must not only identify the disease causing mutations but also quantify the mutation heteroplasmy. Since accurate quantification of mutation heteroplasmy is important in disease prognosis and outcome, MPS needs to provide sufficiently deep coverage for the measurement of heteroplasmy at every single nucleotide position.

A pilot study using MPS technology has been conducted to investigate the possibility and feasibility of mtDNA sequencing using the MPS approach [50, 70]. The entire 16.6-kb mtDNA genome was amplified with two overlapping fragments followed by sequencing with the Illumina Genome Analyzer GAII. The authors demonstrated the ability of MPS to detect reliably mtDNA heteroplasmy down to 5% level and with high sensitivity but poor specificity for variant detection. This method had the advantages of cost effectiveness and fast turnaround time, but cannot detect large mtDNA deletions or mtDNA copy number variation and may include incorrect mtDNA heteroplasmy at certain nucleotide positions. Cui et al. have recently developed a one-step comprehensive analysis of mtDNA by MPS that provides accurate detection of point mutations at every single nucleotide position of the entire mitochondrial genome and determination of large deletions with deletion heteroplasmy and exact breakpoints [71].

Mitochondrial Disorders: Diseases of the Two Genomes

Mitochondrial disorders are a group of complex diseases that can be caused by mutations in both nuclear and mitochondrial genomes. More than 99% of the proteins involved in mitochondrial biogenesis and functions are encoded by an estimated 1,300 nuclear genes. These features of heterogeneous but overlapping clinical presentations and thousands of possible underlying genetic loci make the diagnosis of mitochondrial disorders the most difficult among all inherited disorders [69, 72–75]. The ability of MPS technology to sequence thousands of different genes with deep coverage is an attractive solution to the diagnosis of complex mitochondrial disorders.

A proof-of-concept study was carried out by using a microarray based capture method to enrich the mtDNA genome and 362 nuclear genes related to mitochondrial disorders, followed by sequence analysis on an Illumina Genome Analyzer GAII [40]. Their results showed that more than 94% of the targeted regions were sequenced at a coverage of about 45X for the nuclear genes. Two known mutations in the positive controls were identified correctly. The coverage depth is approximately 4,000X for mtDNA [40].

In this study about 5% of the target regions were not covered, and about 6–10% of the variants identified were novel, which required Sanger confirmation. The authors commented that the sample preparation step was tedious, but the entire procedure, after careful validation and improvement, was likely to be readily adapted for clinical application.

2.1.4 Carrier Testing in Family Members

Of the approximately 7,000 Mendelian genetic disorders, 16% are classified as autosomal recessive disorders. Most autosomal recessive disorders manifest in early childhood with severe disease course and a poor outcome [76, 77]. Preconception screening, coupled with genetic counseling of carriers has helped tremendously in the reduction of the incidence of severe recessive diseases. Thus, an affordable comprehensive preconception screening would be desirable for general population if the assay includes actionable, highly penetrant, and severe recessive mutations. The recently developed target gene capture and MPS approach has made this type of screening assay possible. Bell et al. reported their work on a preconception carrier screening test for 448 severe recessive childhood diseases [51]. A total of 7,717 regions from 437 target genes were enriched by either SureSelect RNA-based in solution capture (Agilent) or microdroplet-based multiplex PCR (RainDance), followed by sequence analysis with Illumina Genome Analyzer GAII or ABI SOLiD.

The average coverage depth was 160X and 93% of regions had >20X coverage. The results demonstrated a specificity of 99.6% and a sensitivity of 94.9%. A study of 104 unrelated individuals revealed an average carrier burden of 2.8 severe pediatric recessive mutations per individual, and the distribution of mutations appeared to be random and pan-ethnic. The authors also commented that about 27% (122 of 460) of the literature-documented mutations are actually common polymorphisms or have been mis-annotated, underscoring the need for better mutation databases.

The authors believed that the combination of two target enrichment methods could provide better detection sensitivity, although at a higher cost. Additionally they commented that sequencing cost could be further decreased with higher test volume. An important issue in the discussion was the availability of informatics support and result interpretation. This remains a major challenge for laboratories looking to implement MPS-based population carrier screening tests in clinical settings in which regulatory guidelines set by CLIA and CAP are required to be followed for certified clinical diagnostic laboratory [51].

2.2 Whole Exome Sequencing

The common procedures for analyzing exome sequencing results include primary analysis to convert image data to base-calling, secondary analysis to align and map sequence reads, and tertiary analysis to annotate the variants. Examples of bioinformatics tools needed for the processing of sequence data can be found at http://seqanswers.com/wiki/How-to/exome_analysis. The bulk of variants detected by MPS are routinely filtered using publicly available SNP databases, the population allele frequency, computational algorithms for function prediction, and disease databases. Comparison of the variants among family members or unrelated patients with similarly defined clinical syndrome is very helpful in narrowing down the genes for further confirmation and investigation.

As described in the examples of target gene sequencing, a large number of patients with clinical diagnosis were left without an identified molecular cause despite sequence analysis of a group of candidate genes responsible for the clinical condition. These results suggested that the disease causing mutations might not be in the genes that were targeted for sequencing or that mutations were not in the coding regions of the targeted genes. Complete exome sequencing (WES) addresses the first problem, but sequencing of the whole human genome, including the non-coding region (WGS), would be needed to solve the second problem.

It has been shown that whole exome sequencing (WES) of a small number of affected, unrelated individuals could potentially be used to identify a causal gene underlying a rare Mendelian monogenic disorder [52]. In a recent report, DNA samples from four affected individuals in three unrelated kindreds with Miller syndrome were subjected to exome capture followed by single-end, 76 bp cycle sequencing [30]. About 40X coverage of 26.6 Mb mappable exome sequence was obtained. After filtering through dbSNP129 and eight HapMap exomes, a single candidate gene, dihydroorotate dehydrogenase (*DHODH*), a key enzyme in the pyrimidine de novo synthesis pathway, was identified. Sanger sequencing confirmed the presence of *DHODH* mutations in three additional families with Miller syndrome [30]. Using this approach, the same group successfully analyzed exomes from ten patients with autosomal dominant Kabuki syndrome. After filtering against SNP databases, candidate genes containing novel variants in all affected individuals were not found. With less stringent filtering criteria, allowing for the presence of modest genetic heterogeneity, multiple candidate genes were found. After phenotypic and genotypic stratification, a gene *MLL2* encoding the trithorax-group histone methyltransferase was identified. Follow-up sequencing detected *MLL2* mutations in 2 of the remaining 3 patients and 26 of 43 additional cases with Kabuki syndrome [52, 53].

The two examples described above are characteristic Mendelian disorders that occurred in multiple unrelated families. The use of exome sequencing identified new disease genes. Recently, this approach enabled researchers to make a clinical diagnosis of a previously undefined disease that altered the treatment in a single child with a life threatening inflammatory bowel disease (IBD) [54]. The male child presented at 15 months with perianal abscesses and proctitis, suggesting an immune defect. However, comprehensive clinical evaluation could not reach a definitive diagnosis. Exome sequencing performed on this individual identified 16,124 variants. After analysis, a novel hemizygous missense alteration that changed a highly conserved cysteine residue to tyrosine in the X-linked inhibitor of apoptosis (*XIAP*) gene was identified as the causative mutation [54]. Functional studies using peripheral blood mononuclear cells (PBMCs) from the patient and normal controls demonstrated defective responsiveness to NOD2 ligands and enhanced apoptosis in the patient's PBMCs, suggesting that the mutation leads to the loss of XIAP activity. Based on the diagnosis, an allogenic transplant of hematopoietic progenitor cells was performed and the child was able to ingest food without recurrence of the gastrointestinal disease. This report demonstrates the power of exome sequencing in molecular diagnosis of a novel disease and the utility of exome sequencing method in a clinical laboratory [54].

2.3 Whole Genome Sequencing

Before WES became commonly used for clinical diagnosis, whole-genome sequencing (WGS) was also proven to be a useful research tool for new gene discovery [9, 78, 79]. Using the ABI SOLiD platform, WGS was performed on the proband of a family with a recessive form of Charcot–Marie–Tooth disease. Compound heterozygous mutations in the *SH3TC2* gene were identified and the mutation were found to segregate with disease in four affected siblings [9].

Similarly, using the ABI SOLiD 4 platform, the disease gene was discovered in a twin brother and sister initially diagnosed with cerebral palsy in early childhood. The WGS revealed an average coverage of 30X and 2,500,000 variants. Extensive filtering resulted in 70 variants. Among them, only three genes were considered to be candidates for an autosomal recessive disorder. One of them was the *SPR* gene, encoding a sepiapterin reductase, responsible for the synthesis of tetrahydrobiopterin, an important cofactor for the biosynthesis of neurotransmitters. Based on the finding, a new therapeutic regimen was administered that led to the improvement of clinical symptoms [11]. This discovery of the gene responsible for DOPA responsive dystonia demonstrated that, similar to WES, the WGS approach can not only provide a molecular diagnosis for patients with rare genetic disorders but also guide effective treatments that would otherwise not have been considered [11, 54].

2.4 *Application of MPS to Prenatal Diagnosis*

Current molecular prenatal diagnosis requires invasive procedures for amniocentesis or chorionic villus sampling. In addition to cost, these procedures have an approximately 0.5% miscarriage risk. Thus, it is desirable to develop a non-invasive procedure for prenatal diagnosis to avoid the risk of fetal loss.

Recently there has been rapid progress in applying MPS techniques to the detection of fetal chromosomal aneuploidies using maternal plasma DNA. These developments have built upon the pioneering work of Denis Lo and his coworkers at The Chinese University of Hong Kong [80], who demonstrated that at the end of first trimester more than 10% of cell-free DNA is from fetal genome. The detection of trisomy 21 aneuploidy from maternal plasma DNA by MPS has been shown to be more reliable than the detection of other aneuploidies such as trisomy 18 and 13 [55]. Subsequently, three large-scale studies involving multiple centers have further confirmed the clinical validity of this approach [81–83]. These studies established that non-invasive detection of fetal trisomy 21 could be carried out at nearly 100% sensitivity and 98% specificity by multiplexed MPS of maternal plasma DNA. Thus, the implementation of the MPS-based detection method will undoubtedly decrease the risk of fetal loss associated with the screening of high risk pregnancies. However, practical considerations, such as turnaround time, cannot be overlooked. It is likely that an invasive procedure will still be required for follow-up.

As for the application of MPS to the prenatal mutation detection of monogenic disorders, differentiating germline from somatic mutations appears to be challenging. A pilot study to detect fetal β-thalassemia mutations using maternal blood and massively parallel sequencing illustrates the possibility of investigating specific genetic disease loci [84].

3 Target Gene Enrichment Methods

3.1 *Multiplex PCR*

3.1.1 Microfluidic Chip Multiplex PCR

As the most common method for Sanger sequencing, the PCR-based enrichment method has the potential to be tailored for MPS based tests. Simplex PCR is unlikely to be applicable since it is quite labor intensive if the target region is large and thousands of PCR are needed. To generate high throughput target amplicons by PCR, multiple methods have been investigated including multiplex PCR and Fluidigm (South San Francisco, CA) Access Array [85, 86]. In a multiplex PCR, several pairs of primers are mixed together in order to amplify multiple regions of the genome in a single PCR tube under the same cycling condition. The Fluidigm access array utilizes a microfluidic chip, which contains separated minute chambers and thus solves the problems of encountered non-specific amplification in regular multiplex PCR. Simplex PCR is performed in each physically partitioned space without cross interference with the others. Under this setting, up to 48 samples can be mixed individually with up to 10 unique primer pairs. Each amplicon or set of amplicons is maintained separately after amplification. With the fixed capacity, microfluidic chip multiplex PCR can be tailored for a small target gene region for enrichment.

3.1.2 Microdroplet-Based Multiplex PCR

Another solution for performing multiple PCR reactions simultaneously with comparable amplification efficiency is using microdroplet-based multiplex PCR. In this setup, each reaction droplet is a reaction well in isolation with a lipid capsule and contains a small amount of template DNA along with one primer pair and the PCR reagents (polymerase, dNTPs, and buffers) [87]. To generate such reaction droplets, template/reagents droplets and primer droplets are first formed separately as individual libraries and then merged together on a microfluidic chip, which is currently commercialized by RainDance Technologies (Lexington, MA). During the PCR reactions, amplifications are carried out independently in each droplet, mimicking a simplex PCR. Therefore, more uniform enrichment of the target regions can be achieved as compared to the conventional multiplex PCR method.

Nevertheless, due to the same intrinsic characteristics, microdroplet PCR technology also faces several challenges shared with other PCR-based assays, including limited availability of primer design at certain genomic regions and relatively small target size. Other enrichment methods, such as capture-based enrichment methods, may be considered as an alternative when large chromosomal regions are targeted.

3.2 Oligonucleotide Probe Based Capture on a Solid Phase

In the array-based capture assays, DNA probes are synthesized on microarrays, ranging from 60 bases to 90 bases in length, for example in NimbleGen probes [88, 89]. Hybridization is performed directly on solid phase array chips to capture the target regions, followed by extensive washing to remove non-specifically bound DNA. Captured DNA is then eluted from the arrays for MPS library construction. Currently, different commercial companies have extended their array capacity with the size of the targeted regions to ~30 Mb. Although it is more time- and cost-efficient than PCR based enrichment, array based capture has its drawbacks: low through-put and difficulty to scale up.

3.3 Oligonucleotide Probe Based Capture in Solution Phase

Solution capture shares most features of array-based capture, except that it does not require dedicated hybridization instrumentation since the capture hybridization is carried out on a regular thermocycler at either 42 °C (for NimbleGen DNA probe based capture) or 65 °C (for Agilent SureSelect RNA probe based capture). In addition, the solution capture processes can easily be scaled up and automated with robotic liquid handling for simultaneous capture of 96 samples in a 96-well plate set-up. Multiple studies have been performed to compare the performance of DNA probe- and RNA probe-based in-solution capture. The results demonstrated similar capture capability with slightly better uniformity achieved by the DNA probe-based method [90–93]. Nevertheless, one common hurdle shared by all capture-based enrichment is the co-capturing of targeted loci/genes with corresponding pseudogenes. At times it is difficult to retrieve only the target sequence if the regions of interest contain long stretches of highly (>90%) homologous sequences such as the case of pseudogene. This is a problem that may not be resolved by the modification of the probe design and experimental procedures. However, pseudogene sequences may sometimes be filtered out by stringent alignment in post-sequencing data management.

4 Cautionary Tale of Current MPS Tests Offered for Clinical Application

As reviewed above, different gene enrichment methods may have certain limitations and these limitations directly affect the subsequent MPS analysis [94, 95]. In addition, read coverage, sensitivity, specificity, and turnaround time may vary depending on the size of the target genes and the chemistry of the different MPS platforms. To bring this new research technology (a series of new technology, in fact) to the standard required for medical application needs stringent validation. In this section we will review the potential shortcomings of the current MPS methodologies and how they will impact on the application to clinical settings.

4.1 PCR-Based Target Amplification Has Intrinsic Drawbacks

This PCR-based amplification followed by MPS encounters some potential drawbacks, which usually appear to produce uneven coverage of the target regions. Poor amplification of some targets may be due to high GC content or particular DNA structures. The presence of SNPs at the primer sites may also cause allelic dropout. In addition, it is difficult to detect large deletion/insertion events using PCR-based enrichment methods.

4.2 Lack of Sufficient Clinical Validation, Specificity, Sensitivity

Unlike the Sanger method, MPS analyses involve more technical steps, including sample preparation, multiplexing, sequencing, and image detection, setting parameters for variant detection, filtering algorithms and cut off values, as well as evaluation of detection sensitivity and specificity. Each of these steps requires quality control and clinical validation. Depending on the enrichment methods, sequencing chemistry and platforms, as well as analytical tools, individual laboratories may institute different standards and procedures. Therefore, some kind of consensus on standard setting is required.

Review of the published articles revealed that the majority of laboratories used a limited number of pre-selected, known positive samples as controls for validation. For example, 5 control samples were from MFS [46] and only 12 known positive controls were used for CDG validation [48]. This level of validation underestimates the false negative rate and thus these methods may have lower assay sensitivity. The validation procedures should be designed in such a way that the full spectrum of the target regions is examined with a large sample size in two phases; phase I with blinded samples and phase II with known positive controls, in order to fulfil the requirement of a full clinical validation.

4.3 High False Negative Rate Due to Low Coverage

Regardless of the methods used for target gene enrichment, low or no coverage for certain exons is a common problem. This is more common in GC rich regions or regions with particular DNA sequence structures that are susceptible to DNA fragmentation. In addition, different chemistry of sequencing platforms, and the different computational algorithms, may miss particular regions of target sequences. For example, indels in the homopolymer regions are easily missed by pyro-sequencing-based technology, and short tandem repeats (STRs) may be missed by ligation based short reads. The low and no coverage regions are the major cause of false negative results, which in a clinical setting are the least acceptable analytical errors. For a panel containing a small number of genes, these low or no coverage regions may be easily filled by conventional PCR/Sanger sequencing. However, even if the missed regions are known, it may be impossible to fill these gaps by Sanger sequencing if there are a large numbers of low coverage regions. As described in the published papers, 28 out of the 215 coding regions of the CDG study [48] and 20% of the exons in the HSP study were not well covered. Moreover, 34 out of 1,912 amplicons in the XLID study failed amplification [37]. Although there is no standard regarding the minimum coverage required for MPS-based clinical tests, a preliminary cut-off of at least 40 reads from both directions was suggested (TECHGENE forum 2011, Leuven, Belgium). At this cut-off, on average, about 5–10% of the target regions are considered not sufficiently covered. That would mean that about ~9,000–18,000 coding exons are not well covered in the whole exome sequence analysis. A recent analysis on the coverage depth with regard to the detection sensitivity and sequencing errors indicated that a few hundred-fold coverage may be needed for accurate diagnosis in some clinical situations [96].

4.4 Confirmation of Detected Variants to Rule Out False Positives

False positive rate is high in MPS experiments. Typically, about 10–20 variants per gene are detected [97]. This number can be greatly reduced after filtering, depending on the filtering criteria. On average, there may be up to two missense variants per gene. As such, if a panel of 20 genes is analyzed, there will be approximately 20 novel variants that require confirmation by a second method, usually by Sanger sequencing. This verification step is necessary for two reasons – to remove incorrect calls due to experimental error, and to confirm the diagnosis. The confirmation may become burdensome or impossible when the number of genes analyzed increases to 1,000 or 20,000 for the whole exome. For a clinical test, all novel variants with possible clinical significance must be verified before reporting. The confirmation of a large number of novel variants is time-consuming, resulting in a long turnaround time, which is impractical for clinical diagnosis.

4.5 Interpretation of Novel Variants: The Most Challenging Task

The human genome reference sequence has been updated constantly (Homo sapiens GRCh37.2 (hg19) is being used now). Although the majority of variants detected by MPS have been observed previously, a normal healthy individual may carry a number of sequence changes, of which, majority are benign [98, 99].

The population based variant database dbSNP (http://www.ncbi.nlm.nih.gov/projects/SNP/) has documented allele frequencies, but some ethnic specific variants may be underrepresented; such alleles are often confused for disease causing mutations. Bioinformatics tools have been used extensively to help with the interpretation of novel variants [100, 101]. However, none of these tools have been validated for clinical use and most of the commonly used algorithms have very high false-positive and false-negative rates [102]. Another issue is that a high number of sequencing errors may be mistaken as de novo events. Family-based sequencing has therefore been proposed to aid in the gene discovery [98, 103, 104]. In chapter 4 of this book, another approach using protein structure information is reviewed for its potential use in MPS.

5 Discussion

In the field of clinical medical genetics, the advancement of molecular diagnosis for detecting inherited genetic defects usually accompanies the development of new technologies. These novel technologies often lead to the rapid discovery of new genes and new diseases, leading to further improvement in molecular diagnosis, patient care, and management [20, 105–107].

Since the development of the Sanger dideoxynucleotide-terminator sequencing method, it gradually became, and has remained, the gold standard for clinical molecular diagnostics, due to its accuracy in detecting small genetic variants. The diagnosis of monogenic genetic disorders using Sanger sequencing usually has a reasonable turnaround time of approximately 6–8 weeks depending on the size and complexity of the gene interrogated. However, the laborious workflow, error-prone nature, and high operational cost make it unattractive in the new genomic era, especially when sequencing large genomic regions are frequently involved.

The sequencing of James Watson's genome in 2008 was the first "personal genome" accomplished by using MPS. The era of genomic medicine has arrived and MPS will be a key feature, which will evolve with reducing price and increasing throughput [7]. The methodology developed at the human genome sequencing center (HGSC) at the Baylor College of Medicine has since paved the road to personalized genome re-sequencing for disease gene discovery, risk factor assessment, and MPS-based molecular diagnoses in a clinical setting.

However, there are growing pains as we move forward with these novel technologies. The biggest challenge is the interpretation of the enormous amount of genomic data and the establishment of genotype and phenotype correlations. There are about 3,000,000 single nucleotide polymorphisms in each individual, and everyone is a carrier of some recessive diseases [108]. The complexity posed by these technologies is illustrated by the recent comparative genomic analysis of an individual using different platforms, only about 60% of the variants called are likely to be real [109]. It is clear that, without careful validation, MPS at its present shape cannot be reliably and widely used in clinical setting for diagnosis of genetic diseases.

Another aspect of the complexity is establishing the physiological and pathological significance of a genetic variant in an individual's genome. For example, a recent newborn screening study revealed that the pathogenic status of the previously identified c.1436C>T (p.P479L) mutation of the *CPT1A* gene is debatable due to its high frequency and low penetrance in the Inuit population [110, 111].

The sequencing errors generated during an MPS run may not be a big problem for the purpose of gene discovery and other research applications, as these studies often involve multiple family members or multiple unrelated pedigrees with the same disease, which could be used for internal validation [52, 53]. However, when clinical molecular diagnosis is to be made for a sample of single patient with uncertain disease diagnosis, the accuracy of the sequencing results becomes very critical and crucial [18, 54, 112, 113]. The relatively high false positive findings demonstrated by different groups in this review need to be resolved before the MPS approach can make widespread impact on patient care.

The use of MPS technology is not a perfect solution to all genetic problems. Similar to other molecular technologies, there are limitations such as the inability to detect large, complex genomic structural rearrangements [95]. It also has difficulty in distinguishing active gene from pseudogene or highly homologous genomic regions. Other technologies, including aCGH and Sanger sequencing, are presently required to complement these shortcomings in order to detect the full spectrum of mutations responsible for genetic disorders. In light of these practical issues in the routine application of MPS to clinical diagnosis and patient care MPS will likely require further improvements in both accuracy and standardization [18].

6 Conclusion

The studies reviewed here demonstrate promising results in the development of MPS technologies for clinical applications [20, 114, 115]. The adaptation of MPS approaches to clinical diagnostics has been an on-going active pursuit. Continual improvement of the accuracy of sequencing chemistries, computational algorithms for alignment, bioinformatics analytical tools, and the improved methods for the interpretation of variants will make MPS a primary tool of clinical laboratories. Although finding the causal change for a simplex case is in fact a difficult undertaking, it will become routine [106]. Many hurdles are expected to be resolved in the near future [18, 94, 104, 116]. There is every reason to expect that reduction in cost and improvement in accuracy and speed will be seen in the coming years.

Acknowledgement This work is partially supported by a grant from Muscular Dystrophy Association (Research #175369) to Dr. Lee-Jun Wong.

References

1. Sanger F, Nicklen S, Coulson AR (1977) DNA sequencing with chain-terminating inhibitors. Proc Natl Acad Sci USA 74(12):5463–5467
2. Pagliarini DJ, Calvo SE, Chang B, Sheth SA, Vafai SB, Ong S-E et al (2008) A mitochondrial protein compendium elucidates complex I disease biology. Cell 134(1):112–123
3. Calvo S, Jain M, Xie X, Sheth SA, Chang B, Goldberger OA et al (2006) Systematic identification of human mitochondrial disease genes through integrative genomics. Nat Genet 38(5):576–582
4. Thorburn D (2004) Mitochondrial disorders: prevalence, myths and advances. J Inherit Metab Dis 27(3):349–362

5. Scharfe C, Lu HH-S, Neuenburg JK, Allen EA, Li G-C, Klopstock T et al (2009) Mapping gene associations in human mitochondria using clinical disease phenotypes. PLoS Comput Biol 5(4):e1000374
6. Levy S, Sutton G, Ng PC, Feuk L, Halpern AL, Walenz BP et al (2007) The diploid genome sequence of an individual human. PLoS Biol 5(10):e254
7. Wheeler DA, Srinivasan M, Egholm M, Shen Y, Chen L, McGuire A et al (2008) The complete genome of an individual by massively parallel DNA sequencing. Nature 452 (7189):872–876
8. Ashley EA, Butte AJ, Wheeler MT, Chen R, Klein TE, Dewey FE et al (2010) Clinical assessment incorporating a personal genome. Lancet 375(9725):1525–1535
9. Lupski JR, Reid JG, Gonzaga-Jauregui C, Rio Deiros D, Chen DCY, Nazareth L et al (2010) Whole-genome sequencing in a patient with Charcot-Marie-Tooth neuropathy. N Engl J Med 362(13):1181–1191
10. Dewey FE, Chen R, Cordero SP, Ormond KE, Caleshu C, Karczewski KJ et al (2011) Phased whole-genome genetic risk in a family quartet using a major allele reference sequence. PLoS Genet 7(9):e1002280
11. Bainbridge MN, Wiszniewski W, Murdock DR, Friedman J, Gonzaga-Jauregui C, Newsham I et al (2011) Whole-genome sequencing for optimized patient management. Sci Transl Med 3(87):87re3
12. 1000 Genomes Project Consortium (2010) A map of human genome variation from population-scale sequencing. Nature 467(7319):1061–1073
13. Mardis ER (2008) Next-generation DNA sequencing methods. Annu Rev Genomics Hum Genet 9(1):387–402
14. Metzker ML (2010) Sequencing technologies – the next generation. Nat Rev Genet 11 (1):31–46
15. Shendure J, Ji H (2008) Next-generation DNA sequencing. Nat Biotechnol 26 (10):1135–1145
16. Gibbs RA (2011) Bringing genomics and genetics back together. Science 331(6017):548
17. Collins FS (2011) Faces of the genome. Science 331(6017):546
18. Green ED, Guyer MS (2011) Charting a course for genomic medicine from base pairs to bedside. Nature 470(7333):204–213
19. Lander ES (2011) Initial impact of the sequencing of the human genome. Nature 470 (7333):187–197
20. Voelkerding KV, Dames SA, Durtschi JD (2009) Next-generation sequencing: from basic research to diagnostics. Clin Chem 55(4):641–658
21. Margulies M, Egholm M, Altman WE, Attiya S, Bader JS, Bemben LA et al (2005) Genome sequencing in microfabricated high-density picolitre reactors. Nature 437 (7057):376–380
22. Bennett S (2004) Solexa Ltd. Pharmacogenomics 5(4):433–438
23. Bennett ST, Barnes C, Cox A, Davies L, Brown C (2005) Toward the $1000 human genome. Pharmacogenomics 6(4):373–382
24. Shendure J, Porreca GJ, Reppas NB, Lin X, McCutcheon JP, Rosenbaum AM et al (2005) Accurate multiplex polony sequencing of an evolved bacterial genome. Science 309 (5741):1728–1732
25. Rothberg JM, Hinz W, Rearick TM, Schultz J, Mileski W, Davey M et al (2011) An integrated semiconductor device enabling non-optical genome sequencing. Nature 475 (7356):348–352
26. Harris TD, Buzby PR, Babcock H, Beer E, Bowers J, Braslavsky I et al (2008) Single-molecule DNA sequencing of a viral genome. Science 320(5872):106–109
27. Korlach J, Bjornson KP, Chaudhuri BP, Cicero RL, Flusberg BA, Gray JJ et al (2010) Chapter 20: Real-time DNA sequencing from single polymerase molecules. Methods in Enzymology. Academic, pp 431–455

28. Flusberg BA, Webster DR, Lee JH, Travers KJ, Olivares EC, Clark TA et al (2010) Direct detection of DNA methylation during single-molecule, real-time sequencing. Nat Methods 7(6):461–465
29. Turner EH, Ng SB, Nickerson DA, Shendure J (2009) Methods for genomic partitioning. Annu Rev Genomics Hum Genet 10(1):263–284
30. Ng SB, Nickerson DA, Bamshad MJ, Shendure J (2010) Massively parallel sequencing and rare disease. Hum Mol Genet 19(R2):R119–R124
31. Lu C, Tej SS, Luo S, Haudenschild CD, Meyers BC, Green PJ (2005) Elucidation of the small RNA component of the transcriptome. Science 309(5740):1567–1569
32. Axtell MJ, Jan C, Rajagopalan R, Bartel DP (2006) A two-hit trigger for siRNA biogenesis in plants. Cell 127(3):565–577
33. Morin RD, O'Connor MD, Griffith M, Kuchenbauer F, Delaney A, Prabhu A-L et al (2008) Application of massively parallel sequencing to microRNA profiling and discovery in human embryonic stem cells. Genome Res 18(4):610–621
34. Martin JA, Wang Z (2011) Next-generation transcriptome assembly. Nat Rev Genet 12 (10):671–682
35. Hawkins RD, Hon GC, Ren B (2010) Next-generation genomics: an integrative approach. Nat Rev Genet 11(7):476–486
36. Lister R, Pelizzola M, Dowen RH, Hawkins RD, Hon G, Tonti-Filippini J et al (2009) Human DNA methylomes at base resolution show widespread epigenomic differences. Nature 462 (7271):315–322
37. Hu H, Wrogemann K, Kalscheuer V, Tzschach A, Richard H, Haas SA et al (2009) Mutation screening in 86 known X-linked mental retardation genes by droplet-based multiplex PCR and massive parallel sequencing. Hugo J 3(1–4):41–49
38. Gowrisankar S, Lerner-Ellis JP, Cox S, White ET, Manion M, LeVan K et al (2010) Evaluation of second-generation sequencing of 19 dilated cardiomyopathy genes for clinical applications. J Mol Diagn 12(6):818–827
39. Voelkerding KV, Dames S, Durtschi JD (2010) Next generation sequencing for clinical diagnostics-principles and application to targeted resequencing for hypertrophic cardiomyopathy: a paper from the 2009 William Beaumont Hospital Symposium on Molecular Pathology. J Mol Diagn 12(5):539–551
40. Vasta V, Ng S, Turner E, Shendure J, Hahn SH (2009) Next generation sequence analysis for mitochondrial disorders. Genome Med 1(10):100
41. Lubin IM, Caggana M, Constantin C, Gross SJ, Lyon E, Pagon RA et al (2008) Ordering molecular genetic tests and reporting results: practices in laboratory and clinical settings. J Mol Diagn 10(5):459–468
42. Maddalena A, Bale S, Das S, Grody W, Richards S, the ALQAC (2005) Technical standards and guidelines: molecular genetic testing for ultra-rare disorders. Genet Med 7(8):571–583
43. Chen B, Gagnon M, Shahangian S, Anderson NL, Howerton DA, Boone JD (2009) Good laboratory practices for molecular genetic testing for heritable diseases and conditions. MMWR Recomm Rep 58(RR-6):1–37, quiz CE-1-4
44. Vance GH (2011) College of American pathologists proposal for the oversight of laboratory-developed tests. Arch Pathol Lab Med 135(11):1432–1435
45. Chou L-S, Liu CSJ, Boese B, Zhang X, Mao R (2010) DNA sequence capture and enrichment by microarray followed by next-generation sequencing for targeted resequencing: neurofibromatosis type 1 gene as a model. Clin Chem 56(1):62–72
46. Baetens M, Van Laer L, De Leeneer K, Hellemans J, De Schrijver J, Van De Voorde H et al (2011) Applying massive parallel sequencing to molecular diagnosis of Marfan and Loeys-Dietz syndromes. Hum Mutat 32(9):1053–1062
47. Schlipf NA, Schüle R, Klimpe S, Karle KN, Synofzik M, Schicks J et al (2011) Amplicon-based high-throughput pooled sequencing identifies mutations in CYP7B1 and SPG7 in sporadic spastic paraplegia patients. Clin Genet 80(2):148–160

48. Jones MA, Bhide S, Chin E, Ng BG, Rhodenizer D, Zhang VW et al (2011) Targeted polymerase chain reaction-based enrichment and next generation sequencing for diagnostic testing of congenital disorders of glycosylation. Genet Med 13(11):921–932
49. Simpson DA, Clark GR, Alexander S, Silvestri G, Willoughby CE (2011) Molecular diagnosis for heterogeneous genetic diseases with targeted high-throughput DNA sequencing applied to retinitis pigmentosa. J Med Genet 48(3):145–151
50. Tang S, Huang T (2010) Characterization of mitochondrial DNA heteroplasmy using a parallel sequencing system. Biotechniques 48(4):287–296
51. Bell CJ, Dinwiddie DL, Miller NA, Hateley SL, Ganusova EE, Mudge J et al (2011) Carrier testing for severe childhood recessive diseases by next-generation sequencing. Sci Transl Med 3(65):65ra4
52. Ng SB, Buckingham KJ, Lee C, Bigham AW, Tabor HK, Dent KM et al (2009) Exome sequencing identifies the cause of a mendelian disorder. Nat Genet 42(1):30–35
53. Ng SB, Bigham AW, Buckingham KJ, Hannibal MC, McMillin MJ, Gildersleeve HI et al (2010) Exome sequencing identifies MLL2 mutations as a cause of Kabuki syndrome. Nat Genet 42(9):790–793
54. Worthey EA, Mayer AN, Syverson GD, Helbling D, Bonacci BB, Decker B et al (2011) Making a definitive diagnosis: successful clinical application of whole exome sequencing in a child with intractable inflammatory bowel disease. Genet Med 13(3):255–262
55. Chiu RWK, Sun H, Akolekar R, Clouser C, Lee C, McKernan K et al (2010) Maternal plasma DNA analysis with massively parallel sequencing by ligation for noninvasive prenatal diagnosis of trisomy 21. Clin Chem 56(3):459–463
56. Rasmussen SA, Friedman JM (2000) NF1 gene and neurofibromatosis 1. Am J Epidemiol 151 (1):33–40
57. Messiaen LM, Callens T, Mortier G, Beysen D, Vandenbroucke I, Van Roy N et al (2000) Exhaustive mutation analysis of the NF1 gene allows identification of 95% of mutations and reveals a high frequency of unusual splicing defects. Hum Mutat 15(6):541–555
58. Judge DP, Dietz HC (2005) Marfan's syndrome. Lancet 366(9501):1965–1976
59. Robinson PN, Arteaga-Solis E, Baldock C, Collod-Baroud G, Booms P, De Paepe A et al (2006) The molecular genetics of Marfan syndrome and related disorders. J Med Genet 43 (10):769–787
60. Salinas S, Proukakis C, Crosby A, Warner TT (2008) Hereditary spastic paraplegia: clinical features and pathogenetic mechanisms. Lancet Neurol 7(12):1127–1138
61. Zimmerman RS, Cox S, Lakdawala NK, Cirino A, Mancini-DiNardo D, Clark E et al (2010) A novel custom resequencing array for dilated cardiomyopathy. Genet Med 12(5):268–278. doi:10.1097/GIM.0b013e3181d6f7c0
62. Sakai H, Suzuki S, Mizuguchi T, Imoto K, Yamashita Y, Doi H et al (2012) Rapid detection of gene mutations responsible for non-syndromic aortic aneurysm and dissection using two different methods: resequencing microarray technology and next-generation sequencing. Hum Genet 131(4):591–599
63. Jaeken J, Matthijs G (2007) Congenital disorders of glycosylation: a rapidly expanding disease family. Annu Rev Genomics Hum Genet 8(1):261–278
64. Cantagrel V, Lefeber DJ, Ng BG, Guan Z, Silhavy JL, Bielas SL et al (2010) SRD5A3 is required for converting polyprenol to dolichol and is mutated in a congenital glycosylation disorder. Cell 142(2):203–217
65. Ng BG, Sharma V, Sun L, Loh E, Hong W, Tay SKH et al (2011) Identification of the first COG-CDG patient of Indian origin. Mol Genet Metab 102(3):364–367
66. Matthijs G, Schollen E, Bjursell C, Erlandson A, Freeze H, Imtiaz F et al (2000) Mutations in PMM2 that cause congenital disorders of glycosylation, type Ia (CDG-Ia). Hum Mutat 16 (5):386–394
67. Fishman GA (1978) Retinitis pigmentosa: visual loss. Arch Ophthalmol 96(7):1185–1188
68. Ropers HH (2008) Genetics of intellectual disability. Curr Opin Genet Dev 18(3):241–250

69. Wong L-JC (2010) Molecular genetics of mitochondrial disorders. Dev Disabil Res Rev 16 (2):154–162
70. Huang T (2010) Next generation sequencing to characterize mitochondrial genomic DNA heteroplasmy. Current Protocols in Human Genetics. Wiley
71. Cui H, Zhang W, Wong L-JC (2011) Comprehensive molecular analyses of mitochondrial genome by next-generation sequencing 12th International Congress of Human Genetics/ 61st Annual Meeting of The American Society of Human Genetics; 2011; Montreal, Canada, 2011
72. Haas RH, Parikh S, Falk MJ, Saneto RP, Wolf NI, Darin N et al (2008) The in-depth evaluation of suspected mitochondrial disease. Mol Genet Metab 94(1):16–37
73. Wong L-JC, Scaglia F, Graham BH, Craigen WJ (2010) Current molecular diagnostic algorithm for mitochondrial disorders. Mol Genet Metab 100(2):111–117
74. Berardo A, DiMauro S, Hirano M (2010) A diagnostic algorithm for metabolic myopathies. Curr Neurol Neurosci Rep 10(2):118–126
75. DiMauro S, Schon EA (2008) Mitochondrial disorders in the nervous system. Annu Rev Neurosci 31(1):91–123
76. Costa T, Scriver CR, Childs B (1985) The effect of Mendelian disease on human health: a measurement. Am J Med Genet 21(2):231–242
77. Kumar P, Radhakrishnan J, Chowdhary MA, Giampietro PF (2001) Prevalence and patterns of presentation of genetic disorders in a pediatric emergency department. Mayo Clin Proc 76 (8):777–783
78. Boone P, Wiszniewski W, Lupski J (2011) Genomic medicine and neurological disease. Hum Genet 130(1):103–121
79. Biesecker LG (2010) Exome sequencing makes medical genomics a reality. Nat Genet 42 (1):13–14
80. Lo YMD, Corbetta N, Chamberlain PF, Rai V, Sargent IL, Redman CWG et al (1997) Presence of fetal DNA in maternal plasma and serum. Lancet 350(9076):485–487
81. Rossa WKC, Ranjit A, Yama WLZ, Tak YL, Hao S, Chan KCA et al (2011) Non-invasive prenatal assessment of trisomy 21 by multiplexed maternal plasma DNA sequencing: large scale validity study. BMJ 342:c7401
82. Ehrich M, Deciu C, Zwiefelhofer T, Tynan JA, Cagasan L, Tim R et al (2011) Noninvasive detection of fetal trisomy 21 by sequencing of DNA in maternal blood: a study in a clinical setting. Am J Obstet Gynecol 204(3):205.e1–e11
83. Palomaki GE, Kloza EM, Lambert-Messerlian GM, Haddow JE, Neveux LM, Ehrich M et al (2011) DNA sequencing of maternal plasma to detect Down syndrome: an international clinical validation study. Genet Med 13(11):913–920
84. Lo YMD, Chan KCA, Sun H, Chen EZ, Jiang P, Lun FMF et al (2010) Maternal plasma DNA sequencing reveals the genome-wide genetic and mutational profile of the fetus. Sci Transl Med 2(61):61ra91
85. Ottesen EA, Hong JW, Quake SR, Leadbetter JR (2006) Microfluidic digital PCR enables multigene analysis of individual environmental bacteria. Science 314(5804):1464–1467
86. Spurgeon SL, Jones RC, Ramakrishnan R (2008) High throughput gene expression measurement with real time PCR in a microfluidic dynamic array. PLoS One 3(2):e1662
87. Tewhey R, Warner JB, Nakano M, Libby B, Medkova M, David PH et al (2009) Microdroplet-based PCR enrichment for large-scale targeted sequencing. Nat Biotechnol 27(11):1025–1031
88. Okou DT, Steinberg KM, Middle C, Cutler DJ, Albert TJ, Zwick ME (2007) Microarray-based genomic selection for high-throughput resequencing. Nat Methods 4(11):907–909
89. Okou DT, Locke AE, Steinberg KM, Hagen K, Athri P, Shetty AC et al (2009) Combining microarray-based genomic selection (MGS) with the illumina genome analyzer platform to sequence diploid target regions. Ann Hum Genet 73(5):502–513
90. Clark MJ, Chen R, Lam HYK, Karczewski KJ, Chen R, Euskirchen G et al (2011) Performance comparison of exome DNA sequencing technologies. Nat Biotechnol 29(10):908–914

91. Sulonen A-M, Ellonen P, Almusa H, Lepisto M, Eldfors S, Hannula S et al (2011) Comparison of solution-based exome capture methods for next generation sequencing. Genome Biol 12(9):R94
92. Asan, Xu Y, Jiang H, Tyler-Smith C, Xue Y, Jiang T et al (2011) Comprehensive comparison of three commercial human whole-exome capture platforms. Genome Biol 12(9):R95
93. Parla J, Iossifov I, Grabill I, Spector M, Kramer M, McCombie WR (2011) A comparative analysis of exome capture. Genome Biol 12(9):R97
94. Robinson PN, Krawitz P, Mundlos S (2011) Strategies for exome and genome sequence data analysis in disease-gene discovery projects. Clin Genet 80(2):127–132
95. Tucker EJ, Mimaki M, Compton AG, McKenzie M, Ryan MT, Thorburn DR (2012) Next-generation sequencing in molecular diagnosis: NUBPL mutations highlight the challenges of variant detection and interpretation. Hum Mutat 33(2):411–418
96. De Leeneer K, De Schrijver J, Clement L, Baetens M, Lefever S, De Keulenaer S et al (2011) Practical tools to implement massive parallel pyrosequencing of PCR products in next generation molecular diagnostics. PLoS One 6(9):e25531
97. The International HapMap Consortium (2007) A second generation human haplotype map of over 3.1 million SNPs. Nature 449(7164):851–861
98. Vissers LELM, de Ligt J, Gilissen C, Janssen I, Steehouwer M, de Vries P et al (2010) A de novo paradigm for mental retardation. Nat Genet 42(12):1109–1112
99. O'Roak BJ, Deriziotis P, Lee C, Vives L, Schwartz JJ, Girirajan S et al (2011) Exome sequencing in sporadic autism spectrum disorders identifies severe de novo mutations. Nat Genet 43(6):585–589
100. Lindblom A, Robinson PN (2011) Bioinformatics for human genetics: promises and challenges. Hum Mutat 32(5):495–500
101. Medvedev P, Stanciu M, Brudno M (2009) Computational methods for discovering structural variation with next-generation sequencing. Nat Methods 6(11s):S13–S20
102. Wei Q, Wang L, Wang Q, Kruger WD, Dunbrack RL (2010) Testing computational prediction of missense mutation phenotypes: Functional characterization of 204 mutations of human cystathionine beta synthase. Proteins 78(9):2058–2074
103. Roach JC, Glusman G, Smit AFA, Huff CD, Hubley R, Shannon PT et al (2010) Analysis of genetic inheritance in a family quartet by whole-genome sequencing. Science 328 (5978):636–639
104. Robinson P (2010) Whole-exome sequencing for finding de novo mutations in sporadic mental retardation. Genome Biol 11(12):144
105. Ross JS (2011) Next-generation pathology. Am J Clin Pathol 135(5):663–665
106. Lucy Raymond F, Whittaker J, Jenkins L, Lench N, Chitty LS (2010) Molecular prenatal diagnosis: the impact of modern technologies. Prenat Diagn 30(7):674–681
107. Robin NH (2011) Dysmorphology in the era of whole exome sequencing. Curr Opin Pediatr 23(6):579–580
108. Yngvadottir B, MacArthur D, Jin H, Tyler-Smith C (2009) The promise and reality of personal genomics. Genome Biol 10(9):237
109. Lam HYK, Clark MJ, Chen R, Chen R, Natsoulis G, O'Huallachain M et al (2011) Performance comparison of whole-genome sequencing platforms. Nat Biotechnol 30 (1):78–82
110. Greenberg CR, Dilling LA, Thompson GR, Seargeant LE, Haworth JC, Phillips S et al (2009) The paradox of the carnitine palmitoyltransferase type Ia P479L variant in Canadian Aboriginal populations. Mol Genet Metab 96(4):201–207
111. Collins SA, Sinclair G, McIntosh S, Bamforth F, Thompson R, Sobol I et al (2010) Carnitine palmitoyltransferase 1A (CPT1A) P479L prevalence in live newborns in Yukon, Northwest Territories, and Nunavut. Mol Genet Metab 101(2–3):200–204
112. Bick D, Dimmock D (2011) Whole exome and whole genome sequencing. Curr Opin Pediatr 23(6):594–600

113. Mayer AN, Dimmock DP, Arca MJ, Bick DP, Verbsky JW, Worthey EA et al (2011) A timely arrival for genomic medicine. Genet Med 13(3):195–196
114. Andrew BS (2011) Exome sequencing: a transformative technology. Lancet Neurol 10 (10):942–946
115. Teer JK, Mullikin JC (2010) Exome sequencing: the sweet spot before whole genomes. Hum Mol Genet 19(R2):R145–R151
116. Jackson L, Pyeritz RE (2011) Molecular technologies open new clinical genetic vistas. Sci Transl Med 3(65):65ps2

Top Curr Chem (2014) 336: 47–66
DOI: 10.1007/128_2012_352

Published online: 24 August 2012

Clinical Applications of the Latest Molecular Diagnostics in Noninvasive Prenatal Diagnosis

K.C. Allen Chan

Abstract The presence of cell-free fetal DNA in the plasma of pregnant women has opened up the possibility of noninvasive prenatal diagnosis. With the advances in molecular techniques of microfluidics and massive parallel sequencing, an increasing number of fetal genetic diseases/conditions can be noninvasively detected using maternal plasma DNA analysis. Remarkably, it has recently been shown that the genome-wide genetic map of an unborn fetus can be constructed through extensive sequencing of maternal plasma DNA. In this chapter the different qualitative and quantitative approaches and related methodology for the analysis of fetal DNA in maternal plasma are discussed.

Keywords Noninvasive prenatal diagnosis · Circulating fetal DNA · Plasma DNA · Aneuploidy detection · Down syndrome · Relative mutation dosage analysis · RMD · Relative haplotype dosage · RHDO

Contents

1 Discovery of Circulating Fetal DNA in Maternal Plasma 48
1.1 Qualitative Analysis of Circulating Fetal DNA in Maternal Plasma 48
1.2 Quantitative Analysis of Fetal DNA in Maternal Plasma by Real-Time Quantitative PCR 50
2 Quantitation of the Mutant Alleles of Circulating Fetal DNA 52
2.1 A Challenge to Real-Time Quantitative PCR 52
2.2 Digital PCR as a Solution 53
3 Aneuploidy Detection 55
3.1 RNA Allelic Ratio Analysis 55
3.2 Epigenetic Allelic Ratio Analyses 57
3.3 Relative Chromosomal Dosage Analysis 57

K.C.A. Chan (✉)
Department of Chemical Pathology, The Chinese University of Hong Kong, Shatin, New Territories, Hong Kong, China
e-mail: allen@cuhk.edu.hk

3.4 Random Sequencing Analysis 59
4 Genome-Wide Analysis of the Fetal Genome 61
5 Conclusion 62
References 63

1 Discovery of Circulating Fetal DNA in Maternal Plasma

Prenatal diagnosis is an established part of obstetric care in most developed countries. Conventional methods for obtaining fetal genetic materials for prenatal diagnosis, include amniocentesis and chorionic villus sampling (CVS), carry a definite risk of inducing miscarriage. In 1997, Dennis Lo et al. discovered that fetal DNA is detectable in the cell-free plasma of pregnant women by showing the presence of Y chromosome sequences in the plasma of pregnant women carrying male fetuses using polymerase chain reaction (PCR) and those chromosome Y sequences were not detectable in the plasma of women carrying female fetuses [1]. This discovery suggests that maternal plasma DNA can be a noninvasive source of fetal genetic materials and has been the prelude to developments of noninvasive prenatal diagnostic tests based on the analysis of fetal DNA in maternal plasma.

1.1 Qualitative Analysis of Circulating Fetal DNA in Maternal Plasma

The early noninvasive diagnostic applications of maternal plasma DNA analysis were based on the detection of paternal-specific sequences. These sequences are only present in the genome of the father but absent in the maternal genome. The detection of such sequences in the maternal plasma would indicate the inheritance of such sequences from the father by the fetus. In the early studies the detection of the Y chromosome-specific sequences in maternal plasma/serum was used for the determination of the fetal gender. The presence of such sequences in the maternal plasma/serum indicates a male fetus and the absence of such sequences is suggestive of a female fetus. In a proof-of-principle study, Lo et al. analyzed the plasma DNA from pregnant women carrying male and female fetuses [1]. Using conventional qualitative PCR targeting the *DYS14* gene on the Y chromosome, they were able to detect Y-specific signals in 24 (80%) of the 30 maternal plasma samples, and in 21 (70%) of the 30 maternal serum samples, from women bearing male fetuses. On the other hand, none of the 13 women carrying female fetuses, and none of the ten non-pregnant control women, had positive results for plasma or serum. These results provide concrete evidence of the presence of fetal DNA in the cell-free plasma/serum of maternal blood. In subsequent studies, circulating fetal DNA was shown to be detectable in maternal plasma at as early as 5 weeks of gestation and the accuracy of fetal gender assessment was shown to be better than 95% even in early gestations [2–4]. Noninvasive prenatal fetal gender assessment was particularly useful for the clinical management of congenital adrenal hyperplasia, as early replacement of corticosteroid during pregnancy can prevent the virilization of the affected female fetus [5].

Soon after the demonstration of the presence of circulating fetal DNA in maternal plasma, this newly discovered phenomenon has been applied to the noninvasive prenatal determination of fetal rhesus D genotype. Rhesus D blood group incompatibility between the fetus and the mother can cause severe hemolytic disease of the fetus or newborn. The prenatal determination of the rhesus D status of a fetus in a rhesus D-negative pregnant woman is useful for guiding the subsequent therapies, e.g., anti-rhesus D immunoglobulins therapy, to avoid the sensitization of the mother. However, any invasive ways for obtaining fetal tissues for rhesus D genotyping/phenotyping would carry risk of immune-sensitizing the mother, thus making the situation worse. The discovery of the presence of fetal DNA in maternal plasma has provided a safe alternative for performing prenatal fetal rhesus D genotyping. In 1998, Lo et al. demonstrated that the *RHD* gene sequences could be detected in the plasma of rhesus D-negative women carrying rhesus D-positive fetuses but not in those carrying rhesus D-negative fetuses [6]. These results were rapidly confirmed by a number of other studies involving much larger number of subjects [7–10]. The accuracies of these reports ranged from 95.7% to 99.8% [7–10]. Because of the robustness of the test, noninvasive RHD genotyping by maternal plasma analysis has been rapidly adopted by many countries as a routine clinical service for rhesus D-negative pregnant women [7, 11, 12].

A similar approach was then applied to the prenatal detection of paternally inherited autosomal dominant conditions, for examples Huntington's disease, myotonic dystrophy, and achondroplasia [13–15]. The qualitative detection of paternal mutations in maternal plasma would indicate the inheritance of such mutations by the fetus and hence indicate that the fetus is affected. However, this approach cannot be applied to the scenarios where the mother is a carrier of the mutation. In those scenarios, both the mutant and the wild-type alleles would be present in the maternal plasma regardless of the mutational status of the fetus. Therefore, qualitative detection of the mutant allele cannot be applied directly to the prenatal diagnosis of autosomal recessive conditions, autosomal dominant conditions where the mother carries the mutation and sex-linked disorders. Furthermore, the interpretation of negative detection of the mutation in a maternal plasma sample is not as straightforward as that for a positive result. Although a negative result usually indicates the absence of such a mutation in the fetal genome, other technical causes also need to be considered. For example, the presence of very low concentration of fetal DNA in the maternal plasma sample or the degradation of the plasma DNA can also result in a false-negative detection of the fetal mutation in the maternal plasma. These analytical causes are particularly important when the sample is taken during early gestation when the fetal-derived DNA is only present at very low concentration in maternal plasma. In this regard, the analysis of a positive control for fetal DNA would be useful for excluding the technical reasons for the false-negative detection of the mutation of clinical interest.

The analysis of sequences on the Y chromosome would be an obvious choice as a fetal DNA marker. However, these Y chromosome-specific fetal DNA markers can only be applied to pregnancies carrying male fetuses. In this regard, other polymorphic markers have been used as positive controls for rhesus D genotyping in some large-scale studies [16, 17]. However, as the polymorphic difference between the fetus and the mother would vary between individual pregnancies, a large number of polymorphic markers would be required so as to achieve a desirable sensitivity for detecting the fetal-specific alleles. Therefore, the development of universal fetal DNA markers that are independent of the fetal gender and polymorphic variations would be very useful for maternal plasma fetal DNA analysis. In this regard, Chim et al. demonstrated that the *SERPINB5* gene is methylated in the blood cells and hypomethylated in the placenta [18]. In non-pregnant individuals, circulating DNA is mainly derived from hematopoietic cells [19] and the circulating fetal DNA in pregnant women is derived from the placenta [18]. Therefore, the placentally derived hypomethylated *SERPINB5* sequences can be used as a fetal DNA marker in maternal plasma and this marker is gender and polymorphism independent. The detection of hypomethylated *SERPINB5* sequences in maternal plasma can be achieved by methylation-specific PCR. However, this procedure involves the bisulfite conversion of DNA which can lead to substantial DNA degradation by up to 93%, thus limiting the accuracy of using hypomethylated *SERPIN5* as a positive control for fetal DNA in maternal plasma. In a subsequent study, Chan et al. showed that the *RASSF1A* gene is unmethylated in the blood cells and hypermethylated in the placenta, the exact reverse pattern of *SERPINB5* [20]. They showed that fetal-specific methylated *RASSF1A* sequences could be detected by a method called methylation-sensitive restriction enzyme-mediated real-time PCR. In this method, the plasma DNA sample is first digested with the methylation-sensitive restriction enzyme, *Bst*UI, such that the non-placentally derived unmethylated *RASSF1A* sequences is degraded – then the uncut methylated *RASSF1A* sequences from the placenta can be detected by real-time PCR. As the enzyme digestion is very specific for the unmethylated sequences, this method is much more sensitive than the bisulfite-based methylation-specific PCR for detecting the low concentration of fetal DNA in maternal plasma. This new universal fetal DNA marker has been shown to be useful for identifying the false-negative results due to the presence of low or absence of fetal DNA in maternal plasma samples [20–22].

1.2 Quantitative Analysis of Fetal DNA in Maternal Plasma by Real-Time Quantitative PCR

In addition to the applications which are based on the detection of the presence and absence of certain fetal-specific DNA sequences in maternal plasma, a broader spectrum of clinical applications has been developed based on quantitative analysis of fetal DNA sequences in maternal plasma. However, conventional molecular genetic methods did not allow very precise quantitation of DNA molecules. This hurdle is overcome by the latest advances in technology.

Soon after the discovery of fetal DNA in the plasma of pregnant women, Lo et al. studied the fetal DNA concentrations in maternal plasma using real-time PCR analysis. By targeting the *SRY* gene they showed that the absolute concentration of fetal DNA in maternal plasma ranged from 3 to 70 genome-equivalents per milliliter plasma (GE/mL), with a mean of 25 GE/mL, in early pregnancies (gestational age 11–17 weeks) and from 77 to 770 GE/mL, with a mean of 290 GE/mL, in late pregnancies (gestational age 37–43 weeks). These figures correspond to mean fractional fetal DNA concentrations of 3.4% and 6.2% in early and late pregnancies, respectively [23]. Recently, using more accurate quantitative methods, the fractional concentrations of fetal DNA has been shown to be higher than the figures reported previously [24]. The median fractional fetal-DNA concentrations measured using digital PCR were 9.7%, 9.0%, and 20.4% for the first, second, and third trimesters, respectively [24]. After delivery of the fetus, fetal DNA is cleared rapidly from the maternal circulation with a half-life of 16.3 min [25]. This observation is very important for the subsequent developments on prenatal diagnostic approaches using maternal plasma DNA analysis as it implies that the fetal DNA from the previous pregnancy would not persist in the present pregnancy to affect adversely the accuracy of maternal plasma DNA-based prenatal diagnostic tests. The mechanisms involved in the clearance of fetal DNA from the circulation have not been completely understood. However, the possibilities of hepatic and renal clearance have been proposed. Regarding the latter possibility, it has been shown that Y chromosome-specific sequences can be detected in the urine of pregnant women carrying male fetuses [26–28]. However, the amount of fetal DNA excreted in urine is relatively low and, by this mechanism alone, cannot account for the rapid clearance of fetal DNA in maternal plasma [29].

Aberrations in the concentration of circulating fetal DNA are observed in various pregnancy-associated conditions [30–32] and in pregnancies involving abnormal fetuses [33–35]. For example, the absolute and fractional fetal DNA concentration is increased in pregnant women suffering from preeclampsia, a pregnancy-associated condition typically occurring during the third trimester, characterized by the presence of hypertension and proteinuria [31, 32]. Interestingly, the elevation in fetal DNA concentration has been shown to precede the onset of clinical symptoms [36] and is hypothesized to be related to the impairment of placental perfusion [37]. Furthermore, it has also been shown that the clearance of fetal DNA is impaired in pregnant women suffering from preeclampsia [38]. The fractional fetal DNA concentration in maternal plasma is also increased in pregnant women carrying fetuses with trisomy 21 (Down syndrome) [33]. However, as the difference between women carrying euploid and trisomy 21 fetuses is relatively small and there is significant overlapping between the levels of the two groups, the measurement of the fetal DNA concentration in maternal plasma cannot provide sufficient discrimination power to serve as a clinical test for Down syndrome [39]. With the development of better analytical platforms, more precise measurement of the quantities of different DNA species in maternal plasma can be achieved. These technical advances have led to the developments of more accurate diagnostic tests for detecting a wide variety of fetal genetic abnormalities. In the following sections the principles of these new clinical diagnostic applications of maternal plasma DNA analysis will be discussed.

2 Quantitation of the Mutant Alleles of Circulating Fetal DNA

2.1 A Challenge to Real-Time Quantitative PCR

The early developments of noninvasive prenatal diagnosis mainly focused on the detection of paternally inherited mutations or sequences in maternal plasma. As discussed above, this approach is particularly useful for the detection of autosomal dominant diseases/conditions in families in which the father carries the mutation. However, in the situation where the mother carries the mutation, both the mutant and wild-type alleles would be detectable in the maternal plasma regardless of whether the fetus has inherited the mutant or the wild-type allele from the mother. This situation would be relevant to the clinical scenarios when both the father and mother are carriers of the mutation in an autosomal recessive condition, and when the mother carries the mutation in an autosomal dominant condition. In such conditions, the quantitative analysis of the mutant and wild-type alleles in the maternal plasma would be required for determining whether the fetus has inherited the mutant or the wild-type allele from the mother. In principle, the allele inherited by the fetus would be present in a higher concentration than the allele not being inherited by the fetus, and the magnitude of the difference would be dependent on the fractional concentration of fetal DNA in the maternal plasma. The relationship between the fetal DNA concentration and the difference in the concentration of the two different maternal alleles for a pregnant woman who is a carrier of a mutant allele and carrying a fetus who is homozygous for the wild-type allele is illustrated in Fig. 1. As this approach is based on the comparison of the relative dosages of the mutant and wild-type alleles in the maternal plasma, this method has been called the relative mutation dosage (RMD) approach [40]. Since the fractional fetal DNA concentration is only around 10% in early pregnancy [24], the concentrations of the mutant and wild-type alleles only differ by approximately 20%. The detection of a 20% difference in the concentration of two different alleles in maternal plasma is technically very challenging. In the early works related to plasma DNA analysis, the quantification of different DNA sequences in plasma was mostly performed using quantitative real-time PCR. For real-time PCR, the quantity of a target sequence in a sample is inferred by the number of PCR cycles required to generate a threshold amount of fluorescence signal. As each PCR cycle would approximately double the amount of PCR products, thereby doubling the amount of fluorescence signal, it is generally accepted that real-time

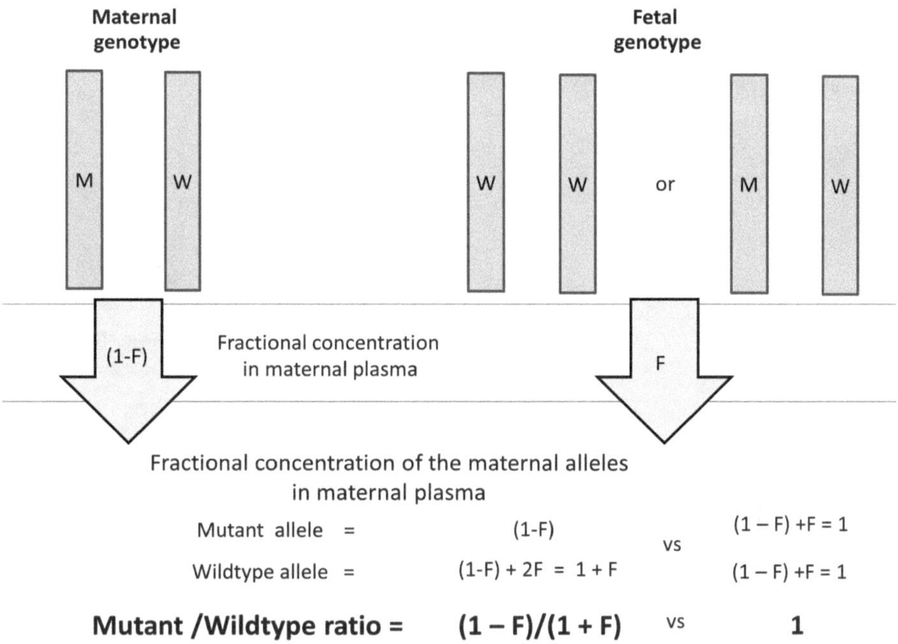

Fig. 1 The principle of the RMD approach. The *M* and *W* rectangles represent the mutant and wild-type alleles, respectively. F represents the fractional concentration of fetal DNA in the maternal plasma

PCR is sufficiently accurate to determine a two-fold difference in the quantities of two targets. In addition, the absolute concentration of DNA in plasma is only around 1,000 GE/mL [24]. Therefore, when a small amount of the DNA target is sampled and analyzed, the concentration of the target sequence in the sample would be subjected to stochastic variation and this would adversely affect the precision of this approach in detecting the quantitative difference between the two alleles [40].

2.2 *Digital PCR as a Solution*

To achieve better accuracy for target sequence quantification, a single molecule counting approach, called digital PCR, was developed [41]. The original protocol of this method involves multiple steps: the DNA sample is first diluted and then distributed into a number of PCR vessels. The concentration of the diluted DNA sample needs to be adjusted so that only approximately half of the vessels would contain a molecule of the target sequence. Under this dilution level, most of the wells would contain either one or zero target molecules. Then real-time PCR can be performed for all of the PCR vessels and the

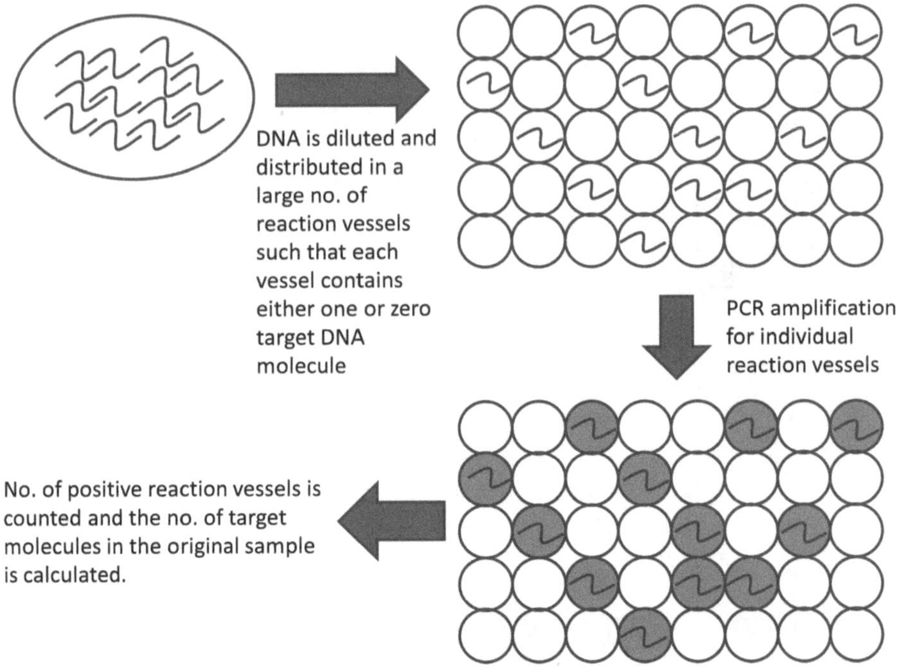

Fig. 2 The principle and workflow of digital PCR analysis

vessels containing a target molecule would give a positive signal. Finally, the number of vessels showing a positive signal can be used to infer the number of target molecules in the original sample. Using digital PCR analysis, the number of target molecules in a DNA sample can be directly counted and this approach would give a more accurate estimation of the target molecules in a sample compared with real-time PCR [24]. The principle and work-flow of digital PCR analysis is illustrated in Fig. 2.

The precision of the digital PCR measurement would be affected by stochastic variation, and magnitude of stochastic variation is dependent on the number of target molecules being analyzed. The higher the number of molecules analyzed, the less the stochastic variation would be affecting the measurement. For RMD analysis, the number of target molecules needing to be analyzed to give accurate results is dependent on the quantitative difference in the concentrations of the mutant and wild-type alleles which in turn depends on the fractional concentration of fetal DNA in the maternal plasma sample [24]. In this regard, Lun et al. performed a computer simulation analysis to estimate the number of target molecules needing to be analyzed so as to achieve a 99% accuracy in the RMD analysis. When the fractional fetal DNA concentration in a plasma sample is 20%, approximately 1,000 molecules would need to be analyzed. This figure increases to 4,000 when the fractional fetal DNA concentration is dropped to 10%, and further increases to

16,000 when the fractional fetal DNA concentration is dropped to 5% [40]. In early pregnancy, the absolute concentration of total DNA in the maternal plasma is approximately 1,000 GE/mL and the fractional fetal DNA concentration ranges from 5% to 10% [23, 24]. Therefore a relatively large volume of maternal blood (>20 mL) would need to be collected so as to provide a sufficient amount of DNA molecules for RMD analysis (also to account for the loss during the different analytical steps). A further refinement of this technique which would allow a more efficient use of plasma DNA and requiring a much smaller volume of maternal plasma will be discussed in a later section.

3 Aneuploidy Detection

To determine whether the fetus is suffering from Down syndrome (trisomy 21) is the most common reason for prenatal diagnosis in a pregnant woman. The development of noninvasive methods for detecting Down syndrome and other chromosomal aneuploidies is therefore a long-sought goal for scientists working in this field. However, the noninvasive detection of fetal Down syndrome is technically challenging because the aberration to be detected is a quantitative change in chromosome 21 dosage rather than the presence or absence of certain sequences in the fetal genome. Here the different approaches developed for noninvasive prenatal diagnosis of Down syndrome are discussed.

3.1 RNA Allelic Ratio Analysis

As fetal DNA is present at a low concentration in maternal plasma, together with a high background of DNA derived from the mother, the detection of chromosomal dosage aberration associated with a trisomy 21 fetus is technically challenging. One possible solution for overcoming this problem would be the selective analysis of the fetal-specific RNA transcripts in maternal plasma. In this regard, studies have demonstrated that the fetal-specific nucleic acids in the maternal circulation are indeed derived from the placenta whereas the maternal-derived nucleic acids are mainly derived from hematopoietic cells [18–20, 42]. Therefore the fetal-specific RNA transcripts can be identified through the comparison between the expression profiles of placentas and maternal blood cells [42]. The transcripts expressed in the placenta but not by the hematopoietic cells would be fetal-specific and the analysis of these transcripts may reveal the genetic composition of the fetus. Among the different fetal-specific transcripts located on chromosome 21, *PLAC4* has the highest expression level in the placenta and, hence, it is used as a target for developing the diagnostic approach depending on the RNA allelic ratio analysis [43]. The principle of this approach is illustrated in Fig. 3. This approach is

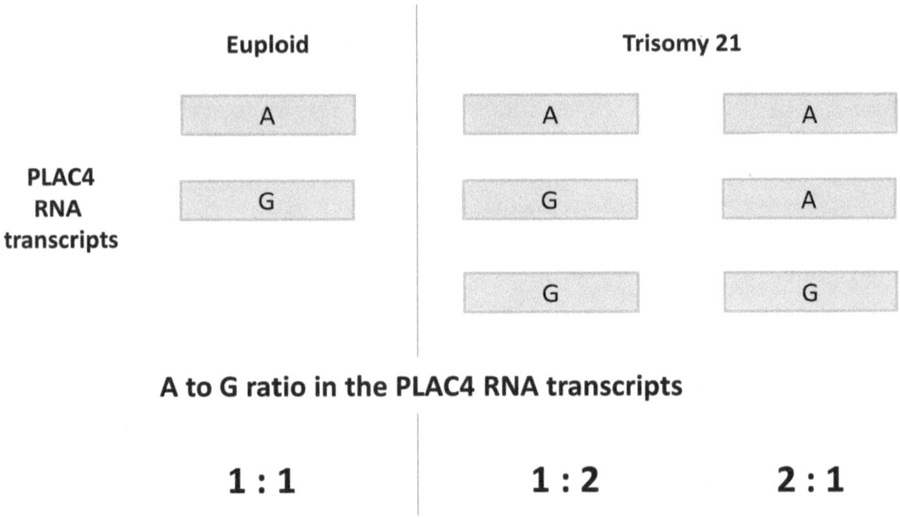

Fig. 3 The principle of the RNA allelic ratio approach

applicable when the fetus is heterozygous at an SNP located on the *PLAC4* gene. If the fetus is euploid (having two copies of chromosome 21), the two alleles located on the two different chromosomes 21 would be present at the same concentration. However, if the fetus is having an additional dosage of chromosome 21, the two alleles would be present at different concentrations. The allele on the extra chromosome would be present at a concentration twice as high as that of the other allele. Therefore, the presence of a dosage imbalance between the two fetal alleles for *PLAC4* transcripts in maternal plasma would indicate the presence of a trisomy 21 fetus. In the initial study, the allelic dosage of *PLAC4* was determined using mass spectrometer [43]. In a cohort of 57 pregnant women with a mean gestational age of 14 weeks, the RNA allelic ratio analysis correctly identified 90% of the trisomy 21 cases and 96.5% of the euploid cases. The accuracy of this approach was subsequently shown to be affected by the plasma concentration of the fetal-specific *PLAC4* transcripts. Hence the measurement of the *PLAC4* transcripts in the maternal plasma would be usefully incorporated into the analysis of the RNA allelic ratio analysis [44]. In a subsequent study it was shown that the *PLAC4* allelic ratio can be more accurately determined by digital PCR analysis and this can further improve the accuracy of the prenatal diagnosis of this approach [45]. Despite having a reasonable diagnostic accuracy, the RNA allelic ratio approach is only applicable to pregnancies in which the fetus is heterozygous for the *PLAC4* gene. Using only one SNP locus on the *PLAC4* gene, as in the original report, this test is applicable to less than 50% of pregnancies. Although increasing the number of SNP loci may potentially improve the coverage of this test, the identification of additional SNP loci in genes that are specifically expressed by the placenta but not the hematopoietic cells is difficult.

3.2 Epigenetic Allelic Ratio Analyses

Similar to the analysis of fetal-specific RNA, DNA exhibiting a fetal-specific methylation pattern can also be used for prenatal detection of fetal aneuploidies. In this regard, Tong et al. reported that the promoter of the *SERPINB5* gene, which is located on chromosome 18, is unmethylated in the placenta and hypermethylated in the blood cells. Using methylation-specific PCR, fetal-specific unmethylated *SERPINB5* sequences can be selectively amplified from the maternal plasma. Using mass spectrometry analysis, the allelic ratio for an SNP on the unmethylated *SERPINB5* sequences can be determined [46]. The presence of allelic imbalance in the fetal-specific unmethylated *SERPINB5* sequences would indicate the presence of an additional chromosome 18 in the fetus and suggests that the fetus is affected by trisomy 18. As this method involves the analysis of allelic ratio in sequences with fetal-specific epigenetic features, it is called the epigenetic allelic ratio (EAR) approach [46]. However, similar to the RNA allelic ratio approach, the EAR analysis is only applicable to those pregnancies in which the fetus is heterozygous for the SNP on the fetal-specific gene and this has limited the practicality of applying these methods as routine clinical tests for the prenatal detection of fetal aneuploidies.

3.3 Relative Chromosomal Dosage Analysis

For RNA allelic ratio and EAR analyses, the aberration in chromosome dosage is inferred from the imbalance in the allelic ratio at an SNP located on the trisomic chromosome when the fetal-specific DNA/RNA sequences in maternal plasma are analyzed. Therefore these methods are only applicable to a subset of pregnancies in which the fetus is heterozygous at the target gene locus. Alternatively, the chromosome dosage of a potentially aneuploid chromosome (the target chromosome) can be compared with another chromosome (the reference chromosome) to determine whether the chromosome dosage of the target chromosome is increased in maternal plasma and this is called the relative chromosome dosage (RCD) analysis. There are two possible ways of performing the chromosome dosage analysis. One way is to focus on the fetal-specific sequences only as in the RNA allelic ratio and EAR approaches. As the trisomy fetuses would have a 50% increase in the chromosome dosage of the target chromosome compared to euploid fetuses, the precision required for detecting the chromosome dosage aberration is less stringent. However, this method requires the identification of fetal-specific markers on the target chromosomes. Alternatively, the chromosome dosage of total DNA in maternal plasma can be determined regardless of whether the DNA is originated from the mother or the fetus. The advantage of this method is that it does not require any fetal-specific DNA marker. However, as only a small proportion of DNA in maternal plasma comes from the fetus, the perturbation in chromosome

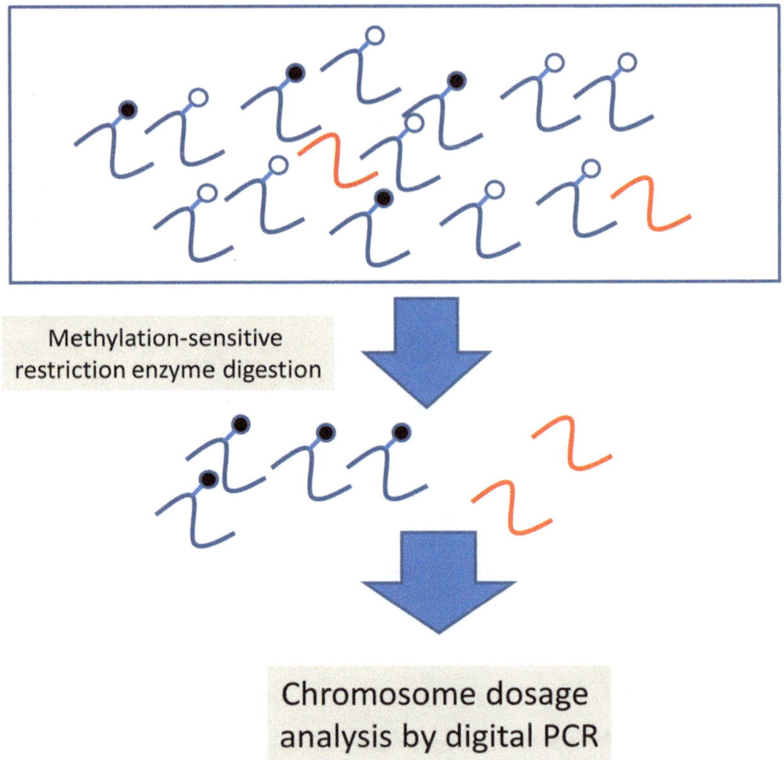

Fig. 4 The principle of the epigenetic–genetic allelic ratio approach. The blue DNA fragments are the target sequences (*HLCS*) on the potentially aneuploid chromosome (chr 21). The red DNA fragments are the fetal-specific sequences on the Y chromosome. The open and close circles indicate unmethylated and methylated status of the DNA fragments, respectively. The methylated fragments are derived from the fetus and the unmethylated fragments are derived from the mother

dosage caused by a trisomy fetus is very small. Therefore the measurement of the chromosome dosage needs to be very precise so as to achieve a good diagnostic accuracy.

Regarding the former approach, Tong et al. developed the epigenetic–genetic chromosome dosage (EAR) analysis and its principle is illustrated in Fig. 4. In this method they used the placentally derived methylated *HLCS* sequences as the fetal-specific targets to determine the chromosome 21 dosage in maternal plasma because the *HLCS* gene is located on chromosome 21 and it is methylated in the placenta and unmethylated in the blood cells [47]. By digesting the maternal plasma DNA with methylation-sensitive restriction enzyme, the unmethylated *HLCS* sequences would be degraded whereas the methylated *HLCS* sequences would remain intact. Then the placentally derived methylated *HLCS* sequences can be quantified. At the same time, a target on another chromosome can be quantified as the reference. One possible reference would be a sequence on chromosome Y.

However, this would only be applicable to the pregnancies with a male fetus. In the proof-of-principle study they used an SNP polymorphism located on the chromosome 14 for determining the dosage of the reference chromosome. Using this method, they demonstrated that the pregnancies involving a trisomy 21 fetus would have an increased chromosome dosage when compared with the pregnancies with a euploid fetus [47].

In addition to using fetal-specific targets for determining chromosome dosage, chromosome dosage analysis can also be performed on the total circulating DNA in maternal plasma regardless of whether they are derived from the mother or the fetus. When the total DNA is analyzed, the degree of increment in chromosome dosage in maternal plasma for pregnant women carrying a trisomy fetus is dependent on the fractional fetal DNA concentration (f) in maternal plasma and is equal to $f/2$. For example, if 10% of the maternal plasma DNA is derived from the fetus, a 5% of increase in chromosome 21 dosage would be observed in the plasma of a pregnant woman carrying a trisomy 21 fetus. As the perturbation in the chromosome dosage is small, the quantification of the targets on different chromosomes needs to be very precise so as to achieve good diagnostic accuracy. In this regard, Lo et al. demonstrated that chromosome dosage analysis can be accurately determined using digital PCR. Using computer simulation analysis, they have established the mathematical relationship between the number of plasma DNA molecules required for an RCD analysis and the fractional fetal DNA concentration when single molecule analysis is to be used for the prenatal diagnosis of fetal aneuploidies [45]. For example, they estimated that approximately 100,000 molecules need to be analyzed if fetal DNA only accounts for 5% of the total maternal plasma DNA so as to achieve a diagnostic accuracy of 98%. However, to analyze 100,000 target molecules is practically difficult when a single target is used. This is because, in early pregnancy, the concentration of DNA in maternal plasma is only around 1,000 GE/mL. One possible solution to this limitation is to analyze multiple targets on the target chromosome and the reference chromosomes. However, the difference in the efficiencies in detecting and quantifying different targets is a potential source of analytical errors.

3.4 Random Sequencing Analysis

In 2008, two groups simultaneously reported the application of massively parallel sequencing for the noninvasive prenatal diagnosis of Down syndrome [48, 49]. The principle of this approach (Fig. 5) is similar to the RCD approach described above. However, instead of counting predefined targets on the potentially aneuploid chromosome and the reference chromosome(s), DNA sequences from any chromosome would be randomly selected from the maternal plasma and a portion of or the entire DNA fragment would be sequenced. After the sequencing, the sequenced DNA fragments would be aligned to the reference human genome to determine their chromosome of origin. Then the proportion of the sequenced DNA fragments

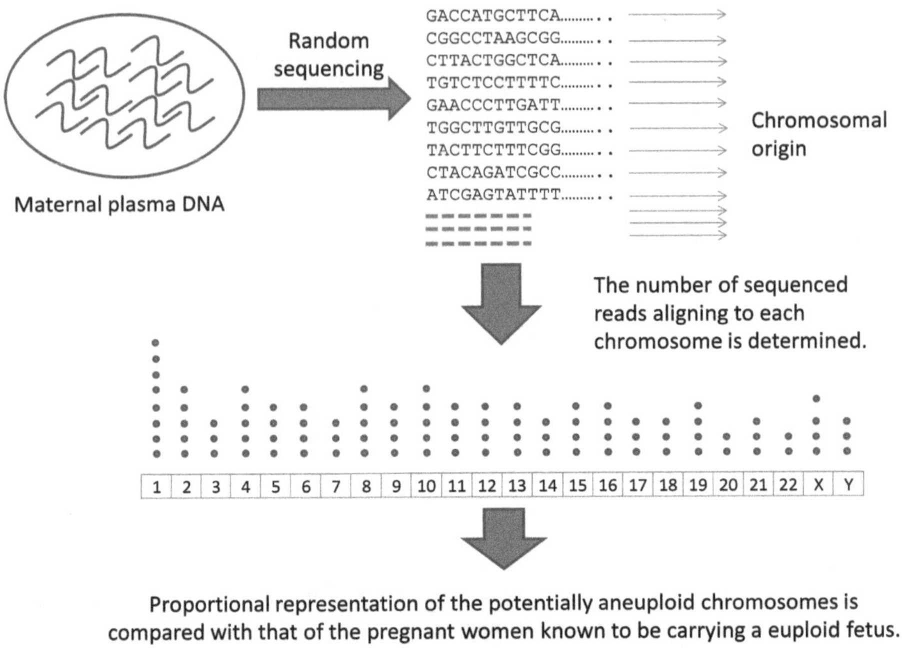

Fig. 5 Noninvasive prenatal analysis for fetal aneuploidy based on random sequencing

from each chromosome would be calculated. As the sequenced fragments are located at different genomic regions, the exact copy number of the target and the reference chromosomes in the plasma sample cannot be determined and the proportion of sequenced fragments aligning to a chromosome would be proportional to the size of that chromosome. When a pregnant woman is carrying a trisomy fetus, the proportion of DNA fragments in maternal plasma originating from the trisomic chromosome would increase, and the magnitude of increment is proportional to the fractional fetal DNA concentration in the maternal plasma as discussed above. The proportional chromosome representation of the tested subject can then be compared with the data from a reference group which consists of pregnant women known to be carrying euploid fetuses to determine whether it is significantly increased. In contrast to all the approaches discussed above, the random sequencing method is applicable to all pregnant women regardless of the gender and genetic polymorphisms of the fetus they are carrying. The precision of measuring the chromosome dosage by this method is dependent on the number of molecules sequenced. In the original study the coefficient of variation in the measurement of proportional representation of chromosome 21 in maternal plasma is only 0.5% and all the 28 cases (14 euploid cases and 14 trisomy 21 cases) were correctly classified [48]. The accuracy of this approach was soon validated in several large-scale clinical studies [50–52]. All of these validation studies reported a sensitivity of 100% for the detection of Down syndrome cases and the specificities of these studies ranged from 98% to 100%. As the precision of measuring the chromosome dosage in plasma is dependent on the amount of maternal plasma DNA fragments being sequenced, it is reasonable to expect that the accuracy of this method can be further improved by sequencing more maternal plasma DNA fragments for each case. Soon after the demonstration of the feasibility of this approach in detecting trisomy 21, this method was also shown to be able to detect other fetal aneuploidies, including trisomy 13 and 18 [53, 54]. Moreover, this method is also sensitive enough to detect the gain of only a part of a chromosome resulting from imbalanced translocation [55].

4 Genome-Wide Analysis of the Fetal Genome

The development of massively parallel sequencing technology has allowed the detailed analysis of circulating DNA in maternal plasma to an unprecedented resolution. Through the analysis of a large amount of DNA molecules, the fetal chromosome dosage can be precisely determined, thus allowing accurate prenatal detection of fetal aneuploidies. To explore the limit of this technology, Lo et al. investigated whether it would be possible to construct the genome-wide genetic map of a fetus through sequencing the circulating DNA of the mother [56]. This task is extremely challenging for several reasons. First, the fetal genome is fragmented into small pieces of less than 200 bp in maternal plasma [57]. In other words, over 30 million small fragments would result from one diploid fetal genome. Furthermore, as half of the fetal genome comes from the mother, there is extensive overlapping between the fetal and maternal genomes. Therefore, the selective analysis of the fetal genome would not be possible. To resolve these problems, the fetal genome was deduced using the paternal and maternal genomes as scaffolds instead of being de novo constructed from fetal DNA in the maternal plasma. In the first step, paternal alleles that are absent in the maternal genome are identified. The presence and absence of these paternal alleles in maternal plasma would indicate which paternal alleles are inherited by the fetus. For the analysis of the maternal alleles, the method used is similar to the RMD approach described previously. In principle, the maternal alleles inherited by the fetus would be present in slightly higher concentration than the alleles that are not inherited by the fetus. However, to deduce accurately the fetal inheritance of all maternal SNPs using RMD, an over 4,000-fold genome sequence coverage would be required even if the fractional fetal DNA concentration is as high as 25% [40]. This high fold sequencing coverage is not practical even using massively parallel sequencing. To resolve this problem, the relative haplotype dosage (RHDO) approach was developed [56]. The principle is illustrated in Fig. 6. In this method, the heterozygous SNPs in the maternal genome are phased to determine the two haplotypes of the mother. The alleles on the same maternal haplotype are analyzed together. From the sequencing data of the maternal plasma, the sequenced fragments carrying the alleles on each of the two maternal haplotypes are counted and compared statistically. The haplotype that is present at higher concentration in the maternal plasma is deduced to be the one inherited by the fetus. Using this approach, Lo et al. successfully determined the genome-wide genetic map of a fetus by sequencing the maternal plasma DNA to

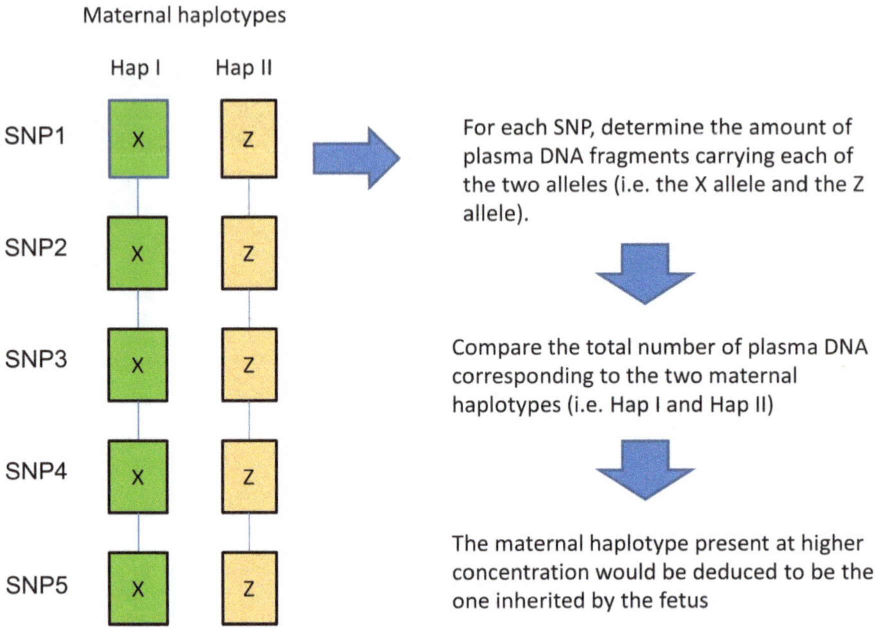

Fig. 6 The principle of the RHDO approach

a coverage of 65-fold [56]. While both the father and mother of this study case are carriers of β-thalassemia mutations, the mutational status of the fetus was also accurately determined.

5 Conclusion

Fifteen years after the discovery of the presence of cell-free fetal DNA in the plasma of pregnant women, a broad spectrum of clinical applications have been developed based on this phenomenon. The demonstration of the noninvasive determination of a genome-wide genetic map of a fetus is particularly intriguing as it indicates that almost all fetal genetic conditions could be noninvasively detected using maternal plasma DNA analysis. In this chapter an overview of how different qualitative and quantitative analyses of maternal plasma DNA can be applied to noninvasive prenatal diagnosis of different fetal genetic diseases and pregnancy-associated complications is given. It is expected that an increasing number of these methods will be adopted for routine clinical use in the near future.

References

1. Lo YMD, Corbetta N, Chamberlain PF et al (1997) Presence of fetal DNA in maternal plasma and serum. Lancet 350:485–487
2. Birch L, English CA, O'Donoghue K et al (2005) Accurate and robust quantification of circulating fetal and total DNA in maternal plasma from 5 to 41 weeks of gestation. Clin Chem 51:312–320
3. Scheffer PG, van der Schoot CE, Page-Christiaens GC et al (2010) Reliability of fetal sex determination using maternal plasma. Obstet Gynecol 115:117–126
4. Wright CF, Burton H (2009) The use of cell-free fetal nucleic acids in maternal blood for non-invasive prenatal diagnosis. Hum Reprod Update 15:139–151
5. Rijnders RJ, van der Schoot CE, Bossers B et al (2001) Fetal sex determination from maternal plasma in pregnancies at risk for congenital adrenal hyperplasia. Obstet Gynecol 98:374–378
6. Lo YMD, Hjelm NM, Fidler C et al (1998) Prenatal diagnosis of fetal RhD status by molecular analysis of maternal plasma. N Engl J Med 339:1734–1738
7. Finning K, Martin P, Summers J et al (2008) Effect of high throughput RHD typing of fetal DNA in maternal plasma on use of anti-RhD immunoglobulin in RhD negative pregnant women: prospective feasibility study. BMJ 336:816–818
8. Muller SP, Bartels I, Stein W et al (2008) The determination of the fetal D status from maternal plasma for decision making on Rh prophylaxis is feasible. Transfusion 48:2292–2301
9. Scheffer PG, van der Schoot CE, Page-Christiaens GC et al (2011) Noninvasive fetal blood group genotyping of rhesus D, c, E and of K in alloimmunised pregnant women: evaluation of a 7-year clinical experience. Int J Obstet Gynaecol 118:1340–1348
10. Van der Schoot CE, Soussan AA, Koelewijn J et al (2006) Non-invasive antenatal RHD typing. Transfus Clin Biol 13:53–57
11. Daniels G, Finning K, Martin P et al (2009) Noninvasive prenatal diagnosis of fetal blood group phenotypes: current practice and future prospects. Prenat Diagn 29:101–107
12. Finning K, Martin P, Daniels G (2004) A clinical service in the UK to predict fetal Rh (Rhesus) D blood group using free fetal DNA in maternal plasma. Ann N Y Acad Sci 1022:119–123
13. Amicucci P, Gennarelli M, Novelli G et al (2000) Prenatal diagnosis of myotonic dystrophy using fetal DNA obtained from maternal plasma. Clin Chem 46:301–302
14. Gonzalez-Gonzalez MC, Trujillo MJ, Rodriguez de Alba M et al (2003) Huntington disease-unaffected fetus diagnosed from maternal plasma using QF-PCR. Prenat Diagn 23:232–234
15. Saito H, Sekizawa A, Morimoto T et al (2000) Prenatal DNA diagnosis of a single-gene disorder from maternal plasma [letter] [in process citation]. Lancet 356:1170
16. Brojer E, Zupanska B, Guz K et al (2005) Noninvasive determination of fetal RHD status by examination of cell-free DNA in maternal plasma. Transfusion 45:1473–1480
17. Zhou L, Thorson JA, Nugent C et al (2005) Noninvasive prenatal RHD genotyping by real-time polymerase chain reaction using plasma from D-negative pregnant women. Am J Obstet Gynecol 193:1966–1971
18. Chim SSC, Tong YK, Chiu RWK et al (2005) Detection of the placental epigenetic signature of the maspin gene in maternal plasma. Proc Natl Acad Sci USA 102:14753–14758
19. Lui YY, Chik KW, Chiu RWK et al (2002) Predominant hematopoietic origin of cell-free DNA in plasma and serum after sex-mismatched bone marrow transplantation. Clin Chem 48:421–427
20. Chan KCA, Ding C, Gerovassili A et al (2006) Hypermethylated RASSF1A in maternal plasma: a universal fetal DNA marker that improves the reliability of noninvasive prenatal diagnosis. Clin Chem 52:2211–2218
21. Li Y, Kazzaz JA, Kellner LH et al (2010) Incorporation of fetal DNA detection assay in a noninvasive RhD diagnostic test. Prenat Diagn 30:1010–1012
22. Zejskova L, Jancuskova T, Kotlabova K et al (2010) Feasibility of fetal-derived hypermethylated RASSF1A sequence quantification in maternal plasma – next step toward reliable non-invasive prenatal diagnostics. Exp Mol Pathol 89:241–247

23. Lo YMD, Tein MS, Lau TK et al (1998) Quantitative analysis of fetal DNA in maternal plasma and serum: implications for noninvasive prenatal diagnosis. Am J Hum Genet 62:768–775
24. Lun FMF, Chiu RWK, Chan KCA et al (2008) Microfluidics digital PCR reveals a higher than expected fraction of fetal DNA in maternal plasma. Clin Chem 54:1664–1672
25. Lo YMD, Zhang J, Leung TN et al (1999) Rapid clearance of fetal DNA from maternal plasma. Am J Hum Genet 64:218–224
26. Al-Yatama MK, Mustafa AS, Ali S et al (2001) Detection of Y chromosome-specific DNA in the plasma and urine of pregnant women using nested polymerase chain reaction. Prenat Diagn 21:399–402
27. Botezatu I, Serdyuk O, Potapova G et al (2000) Genetic analysis of DNA excreted in urine: a new approach for detecting specific genomic DNA sequences from cells dying in an organism. Clin Chem 46:1078–1084
28. Illanes S, Denbow ML, Smith RP et al (2006) Detection of cell-free fetal DNA in maternal urine. Prenat Diagn 26:1216–1218
29. Li Y, Zhong XY, Kang A et al (2003) Inability to detect cell free fetal DNA in the urine of normal pregnant women nor in those affected by preeclampsia associated HELLP syndrome. J Soc Gynecol Investig 10:503–508
30. Illanes S, Parra M, Serra R et al (2009) Increased free fetal DNA levels in early pregnancy plasma of women who subsequently develop preeclampsia and intrauterine growth restriction. Prenat Diagn 29:1118–1122
31. Lo YMD, Leung TN, Tein MS et al (1999) Quantitative abnormalities of fetal DNA in maternal serum in preeclampsia. Clin Chem 45:184–188
32. Tsui DW, Chan KC, Chim SS et al (2007) Quantitative aberrations of hypermethylated RASSF1A gene sequences in maternal plasma in pre-eclampsia. Prenat Diagn 27:1212–1218
33. Lo YMD, Lau TK, Zhang J et al (1999) Increased fetal DNA concentrations in the plasma of pregnant women carrying fetuses with trisomy 21. Clin Chem 45:1747–1751
34. Zhong XY, Burk MR, Troeger C et al (2000) Fetal DNA in maternal plasma is elevated in pregnancies with aneuploid fetuses. Prenat Diagn 20:795–798
35. Zhong XY, Laivuori H, Livingston JC et al (2001) Elevation of both maternal and fetal extracellular circulating deoxyribonucleic acid concentrations in the plasma of pregnant women with preeclampsia. Am J Obstet Gynecol 184:414–419
36. Leung TN, Zhang J, Lau TK et al (2001) Increased maternal plasma fetal DNA concentrations in women who eventually develop preeclampsia. Clin Chem 47:137–139
37. Sifakis S, Zaravinos A, Maiz N et al (2009) First-trimester maternal plasma cell-free fetal DNA and preeclampsia. Am J Obstet Gynecol 201(472):e471–e477
38. Lau TW, Leung TN, Chan LYS et al (2002) Fetal DNA clearance from maternal plasma is impaired in preeclampsia. Clin Chem 48:2141–2146
39. Gerovassili A, Garner C, Nicolaides KH et al (2007) Free fetal DNA in maternal circulation: a potential prognostic marker for chromosomal abnormalities? Prenat Diagn 27:104–110
40. Lun FMF, Tsui NBY, Chan KCA et al (2008) Noninvasive prenatal diagnosis of monogenic diseases by digital size selection and relative mutation dosage on DNA in maternal plasma. Proc Natl Acad Sci USA 105:19920–19925
41. Vogelstein B, Kinzler KW (1999) Digital PCR. Proc Natl Acad Sci USA 96:9236–9241
42. Tsui NBY, Chim SSC, Chiu RWK et al (2004) Systematic micro-array based identification of placental mRNA in maternal plasma: towards non-invasive prenatal gene expression profiling. J Med Genet 41:461–467
43. Lo YMD, Tsui NBY, Chiu RWK et al (2007) Plasma placental RNA allelic ratio permits noninvasive prenatal chromosomal aneuploidy detection. Nat Med 13:218–223
44. Tsui NB, Akolekar R, Chiu RW et al (2010) Synergy of total PLAC4 RNA concentration and measurement of the RNA single-nucleotide polymorphism allelic ratio for the noninvasive prenatal detection of trisomy 21. Clin Chem 56:73–81
45. Lo YMD, Lun FMF, Chan KCA et al (2007) Digital PCR for the molecular detection of fetal chromosomal aneuploidy. Proc Natl Acad Sci USA 104:13116–13121

46. Tong YK, Ding C, Chiu RWK et al (2006) Noninvasive prenatal detection of fetal trisomy 18 by epigenetic allelic ratio analysis in maternal plasma: theoretical and empirical considerations. Clin Chem 52:2194–2202
47. Tong YK, Jin S, Chiu RWK et al (2010) Noninvasive prenatal detection of trisomy 21 by an epigenetic-genetic chromosome-dosage approach. Clin Chem 56:90–98
48. Chiu RWK, Chan KCA, Gao Y et al (2008) Noninvasive prenatal diagnosis of fetal chromosomal aneuploidy by massively parallel genomic sequencing of DNA in maternal plasma. Proc Natl Acad Sci USA 105:20458–20463
49. Fan HC, Blumenfeld YJ, Chitkara U et al (2008) Noninvasive diagnosis of fetal aneuploidy by shotgun sequencing DNA from maternal blood. Proc Natl Acad Sci USA 105:16266–16271
50. Chiu RWK, Akolekar R, Zheng YW et al (2011) Non-invasive prenatal assessment of trisomy 21 by multiplexed maternal plasma DNA sequencing: large scale validity study. BMJ 342: c7401
51. Ehrich M, Deciu C, Zwiefelhofer T et al (2011) Noninvasive detection of fetal trisomy 21 by sequencing of DNA in maternal blood: a study in a clinical setting. Am J Obstet Gynecol 204 (205):e201–e211
52. Sehnert AJ, Rhees B, Comstock D et al (2011) Optimal detection of fetal chromosomal abnormalities by massively parallel DNA sequencing of cell-free fetal DNA from maternal blood. Clin Chem 57:1042–1049
53. Chen EZ, Chiu RW, Sun H et al (2011) Noninvasive prenatal diagnosis of fetal trisomy 18 and trisomy 13 by maternal plasma DNA sequencing. PLoS One 6:e21791
54. Palomaki GE, Deciu C, Kloza EM et al (2012) DNA sequencing of maternal plasma reliably identifies trisomy 18 and trisomy 13 as well as Down syndrome: an international collaborative study. Genet Med 14:296–305
55. Lun FM, Jin YY, Sun H et al (2011) Noninvasive prenatal diagnosis of a case of Down syndrome due to robertsonian translocation by massively parallel sequencing of maternal plasma DNA. Clin Chem 57:917–919
56. Lo YMD, Chan KCA, Sun H et al (2010) Maternal plasma DNA sequencing reveals the genome-wide genetic and mutational profile of the fetus. Sci Transl Med 2:61ra91
57. Chan KCA, Zhang J, Hui AB et al (2004) Size distributions of maternal and fetal DNA in maternal plasma. Clin Chem 50:88–92

Top Curr Chem (2014) 336: 67–98
DOI: 10.1007/128_2012_326

Published online: 20 May 2012

The Role of Protein Structural Analysis in the Next Generation Sequencing Era

Wyatt W. Yue, D. Sean Froese, and Paul E. Brennan

Abstract Proteins are macromolecules that serve a cell's myriad processes and functions in all living organisms via dynamic interactions with other proteins, small molecules and cellular components. Genetic variations in the protein-encoding regions of the human genome account for >85% of all known Mendelian diseases, and play an influential role in shaping complex polygenic diseases. Proteins also serve as the predominant target class for the design of small molecule drugs to modulate their activity. Knowledge of the shape and form of proteins, by means of their three-dimensional structures, is therefore instrumental to understanding their roles in disease and their potentials for drug development. In this chapter we outline, with the wide readership of non-structural biologists in mind, the various experimental and computational methods available for protein structure determination. We summarize how the wealth of structure information, contributed to a large extent by the technological advances in structure determination to date, serves as a useful tool to decipher the molecular basis of genetic variations for disease characterization and diagnosis, particularly in the emerging era of genomic medicine, and becomes an integral component in the modern day approach towards rational drug development.

Keywords Drug development · Genetic diseases · Misfolding · Missense mutations · Mutation analysis · Protein structures · Structure based drug design

Contents

1 Structural Biology in the Post-Genomics Era 68
1.1 Introduction 68
1.2 Methods of Obtaining Structural Information 69

W.W. Yue (✉), D.S. Froese, and P.E. Brennan
Structural Genomics Consortium, University of Oxford, Old Road Campus Research Building, Oxford OX3 7DQ, UK
e-mail: wyatt.yue@sgc.ox.ac.uk

2 Protein Structure Analysis in Understanding Genetic Variations 71
2.1 Studying Diseases in the Next Generation Sequencing Era 71
2.2 Structural Characterization of Missense Variations 72
2.3 The Structural "Rule-Book" Governing Missense Variations 74
3 Protein Structure Analysis in Drug Development 82
3.1 Structural Biology and Target-Centric Drug Development 82
3.2 Early Structural Applications in Lead Optimization 83
3.3 Use of Structures in Target Identification 83
3.4 Use of Structures in Assessing Druggability 85
3.5 Use of Structures in Hit Identification 87
4 Conclusion, Challenges and Future Perspectives 89
4.1 Studying Protein–Protein Interactions 90
4.2 The High Hanging-Fruits of Membrane Protein Structures 90
4.3 Combining Mutation Analysis and Drug Design: Pharmacological Chaperones 91
References 92

1 Structural Biology in the Post-Genomics Era

1.1 Introduction

The past decade has seen an explosion of genome sequences, thanks to the many advances in sequencing technology. These global sequencing efforts have provided us with genetic blueprints for a myriad of organisms in all kingdoms of life. The approach to biomedical research therefore has undergone a radical and dramatic transformation in the post-genomics era. In the emerging era of genomic medicine, it is now possible to sequence completely 3×10^9 base pairs in the human genome for individual patients. We are now tasked with the annotation and description of the plethora of genomic data with regards to biological functions. Although the protein-coding genomic space (exome) is small, where protein-coding exons account for only 1% of the human genome, it represents a majority of the targets for drug development, and 85% of Mendelian diseases are caused by genetic variations in the exomic space. A protein is not merely an "alphabetical" sequence of amino acids, but a macromolecule with three-dimensional (3D) shape and form, capable of performing specialized biological functions in the cell via dynamic interactions with other proteins, small ligands and cellular components. These functional properties depend on a protein's three-dimensional structure, and the field of structural biology is instrumental in directing research towards an understanding of protein function and disease. A large amount of resources have now been put in place, at the disposition of the broad community of non-structural biologists in biomedical research, to exploit the wealth of protein structure information.

In this chapter we aim to provide a brief overview of the current status in protein structure determination, and summarize how protein structure analysis is integral to two active and growing areas of biomedical research, namely understanding genetic variations at a protein level to help disease diagnosis and guiding the development of small molecule therapeutics. Due to the broad subject matter, it is beyond the scope of this chapter to provide an extensive discussion of all significant developments in the ever expanding applications of structural biology. We, however, refer the interested reader to some excellent articles in the relevant sections for more in-depth reviews. We also apologize to all those colleagues whose important work could not be cited, or was cited indirectly, because of space consideration and reference limits.

1.2 Methods of Obtaining Structural Information

1.2.1 Experimental Approaches

As of December 2011, the Protein Structure Database (PDB) contained ~77,700 protein structures in the public domain (http://www.pdb.org). These 3D structures are experimentally derived by methods such as X-ray crystallography, nuclear magnetic resonance (NMR) spectroscopy and electron microscopy (EM). Among them, X-ray crystallography is the dominant structure provider (Fig. 1a), contributing ~87% of total PDB entries. Since the first protein crystal structure in the 1960s (that of myoglobin [1]), the field of protein crystallography has made tremendous technological advances in all stages of the structure determination process. Examples of such development include the use of heterologous systems (e.g. bacteria, baculovirus-infected insect cells, yeast) to recombinantly express proteins in milligram quantities [2], use of fusion tags and automated chromatography platforms to purify proteins, use of robotics in performing nanolitre-scale crystallization experiments [3], improvement of synchrotron and in-house X-ray sources that reduces data collection time and extends resolution limits [4]; and software development to accelerate the *in silico* data processing steps [5]. At present high-resolution crystal structures can often be determined within days of obtaining diffraction-grade crystals.

A second method of structure determination, solution NMR, analyzes resonance assignment derived from short-range inter-proton distances in a protein (Fig. 1b). Compared to crystallography, which requires the protein in a crystalline state, solution NMR benefits from studying the protein in its native form, allowing the observation of protein conformational dynamics and flexibility [6]. NMR provides an alternative route to structure determination, especially for proteins difficult to crystallize, contributing ~11% of total PDB entries. It is also very informative in mapping ligand binding residues, by titration of the ligand onto the protein and analyzing chemical shifts in a heteronuclear single-quantum correlation (HSQC) spectrum. However, solution NMR consumes a considerable amount of radioactive isotope-labelled protein sample and time in the resonance assignment. There is also a size limit for proteins amenable to solution NMR measurement (<30 kDa), although this limit will continue to be pushed back by technological improvements [7].

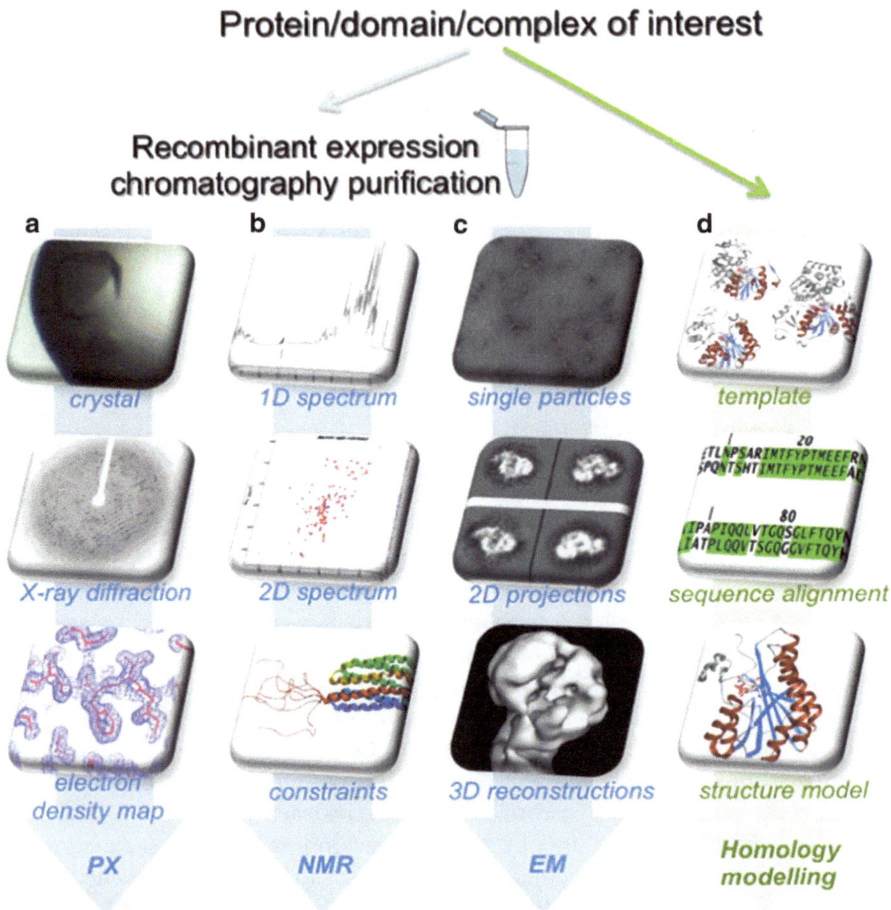

Fig. 1 Methods of protein structure determination. Experimentally protein structures can be determined by protein crystallography (PX), nuclear magnetic resonance (NMR) spectroscopy and electron microscopy (EM). These methods take advantage of recombinant technology that facilitates the heterologous expression and purification of protein domains or complexes. Structure models can also be constructed by homology modeling, if a structure template homologous to the protein of interest is available

Electron microscopy (EM) can determine macromolecular structures at medium to low atomic resolution, using single particle analysis where individual protein molecules are imaged in solution and the 3D structure is reconstructed by back projection of the 2D images (Fig. 1c) [8]. EM is useful in studying multiprotein supramolecular complexes, particularly when combined with crystallography studies of the protein components. This allows the fitting of the individual proteins, of which crystal structures were determined, into the molecular envelope of the intact complex as determined by EM, to understand their relative orientations within the complex. However the use of EM in small molecule development and understanding genetic variations is currently limited by the data resolution restraint.

1.2.2 Homology Modelling

Of the ~30,000 or so gene products predicted for the human genome, only around 15% have been structurally characterized by the experimental methods outlined above. For the many remaining proteins in human and other organisms, computational modeling continues to bridge the gap between known sequences and available structures. The method of comparative or homology modeling allows a structural model to be constructed for a target protein based on its similarity to one or more known structures [9], on the premise that proteins sharing similar sequences fold into similar 3D structures [10]. Today, a number of modeling programs are available (e.g. MODELLER, salilab.org/modeller), some developed as online servers where a sequence-to-structure process can be performed simply by a few clicks. Their popular usage can in part be reflected by the number of available protein models in online repositories such as SWISS-MODEL (swissmodel.expasy.org), Modbase (modbase.compbio.ucsf.edu) and Protein Model Portal (www.proteinmodelportal.org). Due to its popularity, homology modeling has played an influential role in functional annotation and drug discovery for many protein families, e.g. kinases and GPCRs (see examples in [11]).

Common to all modeling tools and servers is an overall four-step procedure (Fig. 1d): (1) given the sequence of the target protein, homologues with known 3D structures are identified; (2) a sequence alignment between the target protein and homologues assigns residue correspondence between sequences; (3) the alignment guides the model building of the target protein, using the homologue structure as template; (4) finally, the constructed model is subjected to refinement and validation of its stereo-chemical properties. In general, the accuracy of homology models depends heavily on the suitability of the template, with higher sequence homology between target and template resulting in less positional errors (as measured by root-mean-square deviations, rmsd, between their corresponding main-chain atoms). In practice, sequence identity cut-offs between 40% [12] and 70% [13] have been used to produce reliable models for understanding protein function and drug discovery (at a 60% identity level rmsd is usually <1 Å). Models derived from lower identity templates ($<30\%$) often have higher main-chain and side-chain errors due to a poor quality sequence alignment with too many position gaps [14].

2 Protein Structure Analysis in Understanding Genetic Variations

2.1 Studying Diseases in the Next Generation Sequencing Era

The recent advent of next generation sequencing (NGS; also known as massively parallel sequencing) has progressed from the time- and cost-consuming Sanger sequencing models to much quicker and cheaper methods [15], and revolutionized our approaches to study the relationship between genotype and disease [16]. Making particular impact has been the use of exome sequencing (i.e. all exons in a genome) to investigate the genetic bases of rare Mendelian disorders with low and sporadic incidence in the population [17]. Its success stems partly from not being technically limited by small patient sample size, a major hurdle with conventional methods of disease gene discovery such as linkage analysis and homozygosity mapping. Today, exome sequencing has led to the discovery of new pathogenic variants and candidate genes for a number of genetic disorders (e.g. Miller syndrome [18], Freeman–Sheldon syndrome [19], Kabuki syndrome [20]), and has also offered opportunities to study complex polygenic diseases (e.g. diabetes, Alzheimer's and heart disease) where susceptibility is affected by multiple genes with complex inheritance patterns.

NGS has therefore accelerated the rate of identifying variants in the human genome. An increasing emphasis is now placed on the effects of these variations on health and disease, although sieving through this huge volume of variant data is a laborious task. Most genetic variations occur at the single nucleotide level, represented as either single nucleotide polymorphisms (SNPs) if they have an incidence of >1% in the genome [21], or as rare variants with <1% occurrence. Rare variants, like SNPs, can be pathogenic (i.e. disease linked; often termed conveniently as mutations) or benign (i.e. not disease linked). Of particular importance to disease diagnostics are those SNPs and rare variants that lead to amino acid substitutions (missense variants) for two reasons. First, the contribution of missense variations to disease is much higher than the summation of all other variant types (e.g. frameshifts, insertions, deletions, splicing, nonsense), with 60–75% of Mendelian disorders caused by amino acid substitutions [22, 23]. Second, while the consequences of most nonsense, frameshifts and insertions/deletions are self-evident (e.g. resulting in truncated proteins), the effects of missense variations on protein function and stability are more subtle and difficult to predict. Structural information at the protein level is therefore needed to understand fully their molecular effects.

2.2 *Structural Characterization of Missense Variations*

While traditionally not a front-line method of analysis, protein structural information has increasingly been incorporated into bioinformatics and *in silico* methods to characterize missense variants and predict their pathogenicity at the molecular level. In the following subsections we outline several approaches of structure-guided investigation of missense variations and the lessons learnt from these studies.

2.2.1 Bioinformatics Predictors

Following the identification of genetic variants, the next indispensable step is to discriminate between pathogenic and benign variations. The sheer volume of genomic data, however, makes it too time-consuming and expensive to characterize every missense variant experimentally. To this end, numerous bioinformatics methods have been developed over the past decade to predict their molecular effects, and thus help prioritize a set of variants to be studied functionally. A number of excellent reviews on the available computational tools have been published recently ([24–26] and references therein for programs described below). Many prediction tools are implemented as online servers, taking an input sequence, and applying various algorithms to sort and score mutations by their pathogenicity. Structure-based algorithms, which identify a structural match to the input sequence and analyze the contributions of the variant amino acid to protein structural properties such as electrostatics, inter-residue contacts, and steric effects [27], are increasingly incorporated into prediction servers. They serve as complementary approaches to the sequence-comparison programs (e.g. SIFT, Panther and PhD-SNP) that are based on the premise that disease-causing mutations are generally concentrated at conserved amino acids with critical roles in protein structure and function [28]. Nowadays, many prediction methods combine both structural information and sequence conservation to improve their prediction performance and accuracy (e.g. nsSNP Analyzer, PolyPhen-1/2, SNAP, SNP&GO and SNPs3D). An emerging trend is to utilize multiple sets of prediction programs and servers to increase confidence in interpreting the predictions, since different algorithms use different information and have their own strengths and weaknesses. Currently there is an urgent need for standard classification as well as unbiased and statistically-relevant comparisons among the various programs, an active area of bioinformatics research [29, 30].

2.2.2 *In Silico* Structural Analysis

Protein structure analysis offers a promising avenue to study the molecular consequences of missense variants, by revealing the atomic environment surrounding the mutation site *in silico*. The most direct approach is by experimental structure determination of the protein in its mutant form if it can be expressed recombinantly and purified. This, however, often proves difficult, in part due to unstable conformations of the mutant proteins that lead to their intracellular degradation. Indeed, three-quarters of disease-associated missense mutations are postulated to destabilize the protein as their primary functional defect [12, 31]. Therefore, although thousands of proteins involved in different biological pathways and functions have been structurally characterized, only a very small proportion of these structures represent proteins inclusive of a disease associated mutation. Recent structure examples falling into this category include the ryanodine receptors RyR1 and RyR2 [32], FGFR2 tyrosine kinase domain [33] and glycogenin GYG1 [34]. As structural determination of mutant proteins often proves intractable, the alternative is to "model" the missense variation onto the wild-type structural environment, by fitting the new amino acid side-chain into the substitution site. The modeled amino acid is inspected visually using molecular graphics software, such as PyMOL (Schrödinger, LLC.), Swiss-PDB viewer (Swiss Institute of Bioinformatics, Basel) and ICM (Molsoft, La Jolla), with particular attention paid to identifying the most acceptable side-chain conformation, from a library of allowed side-chain rotamers, that results in minimal steric clashes. This structure model is then subjected to refinement and energy minimization to yield an overall stabilized conformation.

The available mutant model, either from experimental methods or mutation modeling, can then be analyzed *in silico* to assess the impact of the amino acid substitution on a number of structural properties. These include possible changes in secondary structure elements, solvent accessibility, packing of neighbouring atoms and inter-atomic/inter-protein contacts, many of which can be examined using online tools ([24] and references therein). The *in silico* observations allow hypotheses about the molecular nature of the mutational defects to be made, and subsequently tested using a variety of biophysical and biochemical assay methods. Oligomeric state of the protein, for example, may be assessed by native gel electrophoresis, size-exclusion chromatography (SEC), analytical ultracentrifugation, or dynamic light scattering [35, 36]. Secondary and tertiary structure contents may be assessed by far-UV circular dichroism (CD) [37]. Protein unfolding may be monitored by chemical or thermal denaturing detected with far-UV CD or fluorescence [38]. Functional interactions with protein partners can be determined using co-immunoprecipitation followed by Western blot and SEC [39]; thermodynamics of protein binding to ligand or peptide can be determined via isothermal calorimetry (ITC) or surface plasmon resonance (SPR) [40]. Enzymatic catalysis and Michaelis–Menton kinetics can be measured if an assay specific to the protein of interest is available [33, 41]. The above list is non-exhaustive, as there are many options to examine every aspect of a protein's functional properties in the laboratory. Regardless of the approach(es) chosen, however, it is important to compare the results obtained with the mutant protein against that of wild-type before interpretations are made. It is also important to complement in vitro observations with in vivo studies to comprehend fully the physiological consequences, for example by introducing the variants into the relevant cell lines or genetically engineered animal models.

2.3 The Structural "Rule-Book" Governing Missense Variations

In the following section we review examples illustrating how a structural analysis of the atomic environment surrounding the variant residues, complemented with biochemical and biophysical studies, can be used to attribute deleterious phenotypes to different molecular effects. Together, these examples allow us to formulate a set of "structural rules" to help predict the likely deleterious effects of a missense variation, and can serve as an important toolkit for clinicians and geneticists who need to assess the disease relevance for any newly-identified variations.

2.3.1 Disrupting Protein Fold and Architecture

Phenylketonuria as a Paradigm of Misfolding Diseases

Computational analysis of disease causing variations predicts that ~75% of mutations lead to protein destabilization, while only 7% directly affect biochemical function, suggesting that, for many monogenic diseases, a change in protein stability is the major contributor to disease pathology [12, 31, 42] and giving rise to the concept of misfolding diseases [43]. A classic example of a misfolding disease is phenylketonuria (PKU; OMIM 261600) caused by destabilizing mutations in phenylalanine hydroxylase (PAH). PKU is the most common inborn error of amino acid metabolism (incidence of ~1 in 15,000) with more than 500 deleterious mutations reported, 60% of which are missense mutations scattered across the polypeptide [44]. The majority of mutations result in enzyme forms with reduced stability and a propensity to aggregate, resulting in protein degradation and turnover [45]. To understand how these mutations lead to a misfolded state, the available crystal structures have served as excellent tools to scrutinize the atomic environment of the missense mutation sites [46, 47] and to correlate between genotypes and phenotypes [48–50].

A common cause of destabilizing mutations is a structural perturbation to the protein core by a number of molecular mechanisms, depending on the nature of the original wild type and mutant residues. (1) Mutation of a large buried residue to a small one will create an unfavourable solvent cavity within the core, with larger cavities resulting in greater destabilization [51]. In PAH, mutations of buried phenylalanines (F39L, F55L, F372L), valines (V177A, V190A, V245A) and leucines (L255V, L348V) to smaller residues are commonly found (Fig. 2a). (2) The reverse of the above is also true. Mutations of small residues to large ones require the protein to accommodate bulky side-chains by disturbing the surrounding packing and secondary structure arrangements. Examples in PAH include a number of alanine-to-valine substitutions (A47V, A246V, A259V, A403V) (Fig. 2b). (3) Mutations of non-polar residues within a hydrophobic environment to polar residues may also destabilize a protein because of the thermodynamic penalties incurred on the unbonded polar group. These include mutations of isoleucine to serine (I94S) or threonine (I164T, I174T) or mutations of leucine to serine (L48S, L255S) (Fig. 2c). (4) Finally, mutations of polar and charged side-chains to hydrophobic ones may remove important stabilizing contacts (e.g. electrostatic or hydrogen-bonding interactions). This is especially true of arginines, such as Arg241 (R241C, R241H) and Arg252 (R252G, R252Q, R252W) in PAH (Fig. 2d).

In contrast to core residues, very few protein destabilizing mutations reside on the protein surface, as they can often be substituted with little effect [51]. However, there are exceptions if the mutation disrupts a hydrogen bond or electrostatic interaction at the surface (e.g. D84Y, R176L, R413P mutations in PAH) (Fig. 2e), or if the mutation affects the functional oligomeric state. To this end, PAH forms a tetramer, and mutations that interfere with its tetramerization, e.g. the

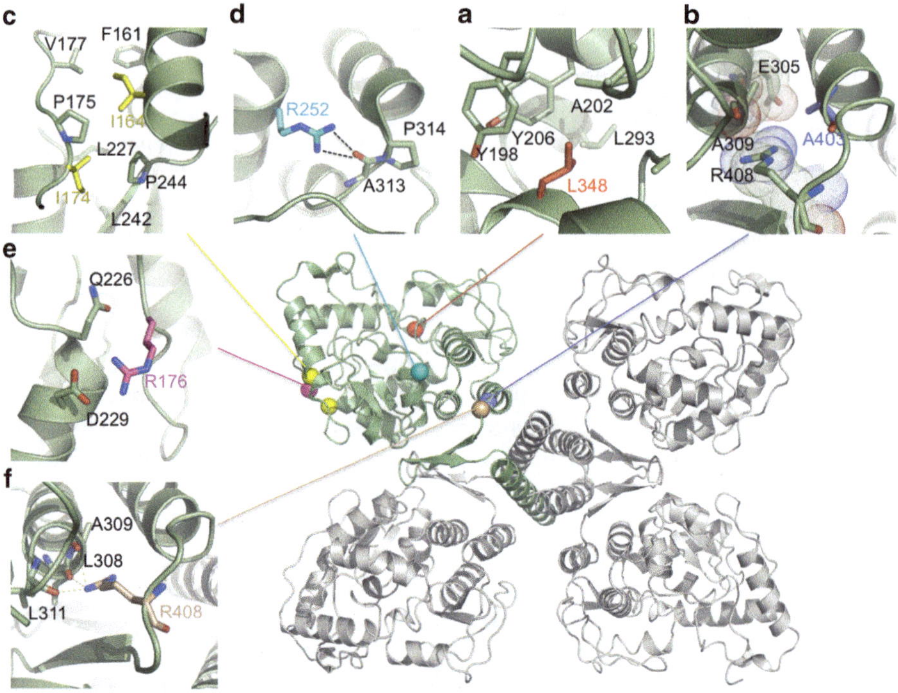

Fig. 2 Structure of human phenylalanine hydroxylase PAH. The tetrameric architecture of PAH (PDB code 2PAH) is shown with one of its monomer subunits coloured in *green*. Six regions of the PAH monomer are highlighted in panels **a–f** to illustrate the different molecular mechanisms that can govern a destabilizing missense mutation. These include (**a**) mutation of larger to smaller residues; (**b**) mutation of smaller to larger residues; (**c**) mutation of nonpolar to polar residues in hydrophobic core; (**d**) mutations of polar to nonpolar residues; (**e**) mutation of surface polar residues; and (**f**) mutations of residues involved in the oligomerization interface

single most common PKU mutation, R408W, which results in the loss of an inter-subunit hydrogen-bond (Fig. 2f), causes improper oligomeric assembly and hence reduces stability [47].

"Special" Residues: Glycine, Proline and Cysteine

Amino acids such as glycine, proline and cysteine often impart certain structural constraints on the protein, and their substitutions can be deleterious. Proline, with its cyclic side-chain, restricts the protein backbone conformations. Therefore, mutations to proline often distort the native backbone conformation, and interrupt the α-helix or β-sheet in which the mutated amino acid resides. The L166P mutation in the DJ-1 protein, located in the middle of helix α7 in its crystal structure (Fig. 3a), is one of the most deleterious missense mutations linked with early onset Parkinson's disease. A combination of NMR, CD and molecular dynamics studies

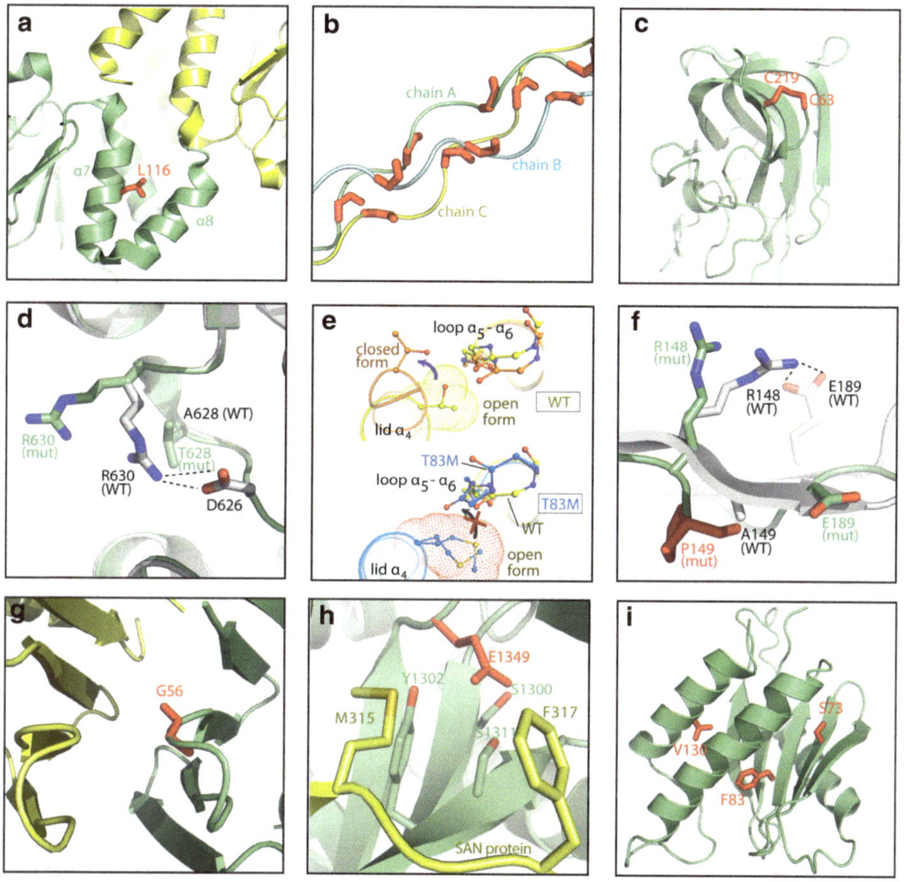

Fig. 3 Defective protein functions due to missense mutations. Where applicable, the site of mutation described in the text is coloured *red*. (**a**) Human DJ-1 protein (PDB code 1PDV). Two monomeric subunits (*green, yellow*) are shown. (**b**) Collagen-like peptide (1CAG) in a triple helix conformation. Glycine residues are shown in *sticks*. (**c**) Structure of Factor VIII C2 domain (1IQD) that is homologous to retinoschisin highlights the highly-conserved disulphide bond (Cys63–Cys219 in retinoschisin). (**d**) Structure of FGRF2 tyrosine kinase domain in wild-type (1GJO, *white*) and A628T mutant (3B2T, *green*). (**e**) Structures of human glycogenin-1 show that the conformational movement of lid α4 in the wild-type (3T7O, *top*) is forbidden in the T83M mutant protein (3RMW, *below*). (**f**) Aldolase B in the wild-type (1QO5, *white*) and A149P mutant protein (1XDM, *green*). (**g**) Spermine synthase (3C6K). The G56S mutation is located at the dimer interface (*yellow, green*). (**h**) Myosin MyoVIIa (*green*) in complex with SAN protein (*yellow*) (3PVL). (**i**) SDELIN protein (1H3Q) with the site of missense mutations disrupting transcription factor binding shown in *red sticks*

have shown that the L166P substitution causes DJ-1 to lose α-helical content and leads to global structural destabilization. Since helices α7 and α8 engage in numerous inter-molecular contacts, the mutant is also incapable of functional dimer formation [37].

With the absence of a side-chain, glycine is the smallest of all amino acids and possesses conformational properties and freedoms inaccessible to other amino acids. Therefore, substitutions from glycine can be debilitating to protein stability and folding. The major structural component of skin, bone and tendons is type I collagen, where two $\alpha 1$ and one $\alpha 2$ protein chains are tightly packed in a heterotrimer. The intermolecular interface is mediated by many Gly-x-y sequence repeats from the three chains, forming a triple helix conformation that is essential to collagen structure and function (Fig. 3b). Many missense mutations substituting a single glycine to larger residues are known to cause the brittle bone disease osteogenesis imperfecta (OMIM 166200), with disease severity dependent upon the size of the mutant amino acid [52]. Another example involves a Gly-to-Asp mutation at the hairpin turn of glycogen phosphorylase, which causes glycogen storage disorder type VI [53].

Cysteine is unique among amino acids in its ability to form inter-residue disulphide bonds that are often critical to maintaining the protein fold. Therefore substitution of a cysteine involved in disulphide bond formation, or to a cysteine that yields an unnatural disulphide bond, may disrupt protein structure. Retinoschisin (RS), a photoreceptor and bipolar cell secreted protein, forms a large disulphide-linked multisubunit complex. At least 25% of the >125 known RS mutations result in the loss or gain of a cysteine and cause X-linked juvenile retinoschisis (OMIM 312700). A combined biochemical and modeling study showed that among the disease causing mutations, C142W and C219R resulted in the breakage of intra-subunit disulphide bonds (Cys110–Cys142 and Cys63–Cys219, respectively) (Fig. 3c), while C59S and C223R abolished an inter-subunit disulphide bond (Cys59–Cys223) [54], hence providing a molecular explanation to how these mutations lead to misfolded protein, defective subunit assembly and aberrant subcellular localization.

2.3.2 Disrupting Protein Functions

While less prevalent than destabilizing mutations, an amino acid substitution can lead to the specific loss, or diminishing, of a protein functional property, such as catalysis, protein–protein interactions, and oligomerization. A number of recent structure examples that are complemented with functional studies are described below.

Affecting Enzyme Catalysis

Mutations in the tyrosine kinase domain (e.g. A628T) of fibroblast growth factor receptor 2 (FGRF2) cause lacrimo-auriculo-dento-digital syndrome (OMIM 149730). Ala628 is a highly conserved residue in the active site catalytic loop. The crystal structure of $FGFR2_{A628T}$ mutant protein reveals that substitution of Ala628 to a more polar and bulky threonine residue alters the configuration of key residues in the active site that are involved in tyrosine substrate binding [33]. For example, the side-chain of Arg630 has been shifted 160° away (Fig. 3d) and cannot coordinate with the substrate. This observation is supported by activity assays showing weakened substrate binding and severely impaired tyrosine kinase activity [33].

A new form of glycogen storage disorder (GSD15; OMIM 613507) has recently been identified with genetic defects in glycogenin (GYG1), a glycosyltransferase that catalyzes the initiation of glycogen synthesis. The complete structural snapshots of GYG1 along its catalytic cycle have been provided by X-ray crystallography and show a substantial "lid" movement that closes the active site for catalysis [34]. The disease-linked mutation T83M incorporates a bulky Met side-chain into the mobile "lid" region and prevents the essential movement, as revealed in the mutant protein structure (Fig. 3e). As a result, the glycosyltransferase activity of $GYG1_{T83M}$ is completely abolished.

Disruption of Quaternary Structure

Hereditary fructose intolerance (OMIM 229600) is caused by mutations in aldolase B, the most prevalent being A149P. The mutant protein structure shows that the A149P substitution disrupts the β-strand element at the mutation site, abolishes a salt-bridge at the adjacent Glu148 residue (Fig. 3f) and also produces a distal effect causing disorder in the 110–129 loop at the dimer–dimer interface [55]. This offers an explanation as to why the mutant protein exists as a solution dimer, and cannot form the homotetramer essential for its catalysis [35]. This study also nicely elucidates the long-range structural perturbations caused by a single amino acid substitution, an observation which would not have been elucidated by modeling a mutant side-chain onto the wild-type structure.

Genetic defects in spermine synthase (SMS), an enzyme converting spermidine to spermine, cause the X-linked disorder Snyder–Robinson Syndrome (OMIM 309583). The crystal structure of SMS reveals that the protein is a homodimer, with the G56S disease mutation lying close to the dimeric interface (Fig. 3g). Any side-chain incorporated at this position is postulated to protrude towards the opposite subunit and disrupt dimer stability, a hypothesis supported by native gel analysis showing the absence of dimer formation in the mutant protein [36].

Disruption of Protein–Protein Interaction

Mutations in the myosin protein MyoVIIa, part of a complex network of proteins in the stereocilia of the inner ear, cause syndromic deaf-blindness (OMIM 276900). A recent structural determination of the MyTH4-FERM tandem domain of MyoVIIa in complex with its protein binding partner Sans reveals that the Glu1349 mutation site on MyoVIIa forms direct interaction with Sans (Fig. 3h), and as a result a single E1349K substitution is responsible for a 20-fold reduction in binding affinity towards Sans, as measured by ITC [40].

Four missense mutations on the SDELIN protein, a subunit of the endoplasmic reticulum Transport Protein Particle complex, are known to cause the X-linked rare bone disorder spondyloepiphyseal dysplasia tarda (OMIM 313400). Three of these mutations (S73L, F83S and V130D) are located in a hydrophobic pocket (Fig. 3i) that is proposed to function as a binding site for transcription factors such as MBP1, PITX1 and SF1, on the basis of the SDELIN crystal structure. Yeast two-hybrid studies have confirmed that these three mutations indeed resulted in a loss of protein–protein interactions [56].

2.3.3 Hot Spot Regions

In addition to visualizing the atomic environment of individual mutation sites, as detailed in Sects. 2.3.1 and 2.3.2, structure analysis can also be employed at the whole protein level, for instance, to map all known variations onto the protein 3D structure and identify "hot spot" regions that harbour a high frequency of missense variations. Hot spot mapping can provide insight into phenotype–genotype relationship of mutations in a 3D structural context, and assist in disease diagnosis, for example, by focusing screening efforts on selected mutation-prone regions instead of over an entire gene, most of which may harbour no known mutations. Hot spot mapping can also help generate new conclusions about protein functions and evolutionary mechanisms such as mutability and selection pressure of different mutations by illustrating which regions of a protein can tolerate amino acid variations and which regions are intolerant. A classic example of hot spot identification is with the most commonly mutated cancer gene, TP53 (p53). In p53, an overwhelming majority of its somatic missense mutations are clustered into a loop-sheet-helix region of the DNA-binding domain (Fig. 4a) [57]. These mutations generally disrupt the DNA binding interface and hence mutant proteins are defective in sequence-specific DNA binding [58].

More recent examples of hot spot mapping can also be found in the literature. Mutations on the ryanodine receptors RyR1 and RyR2 (cf. Sect. 2.2.2) that lead to skeletal muscle disorders are concentrated in a highly basic loop (Fig. 4b) and have been found not to affect protein stability, but rather to disrupt the protein–protein or domain–domain interface [32, 59]. In another example, 15 missense mutation sites on dyskerin, the catalytic subunit of the Box H/ACA ribonucleoprotein particles, have been identified to cause a bone marrow failure called X-linked dyskeratosis congenita (OMIM 305000). The recently determined structure of the yeast homologue Cbf5 reveals that these mutations are all located in a 32-residue N-terminal extension (Fig. 4c) that forms an additional layer to the well-characterized RNA-binding PUA fold, a structural feature not found in archaea, and may function in protein–protein binding [60]. Within our group, we have mapped 55 known missense mutations causing fumarate hydratase deficiency (OMIM 606812) onto its human protein structure [61] and identified two hot spot regions, one clustering around the active site and the other affecting intra- and inter-subunit interactions. To aid further investigation by interested doctors/researchers, the online version of

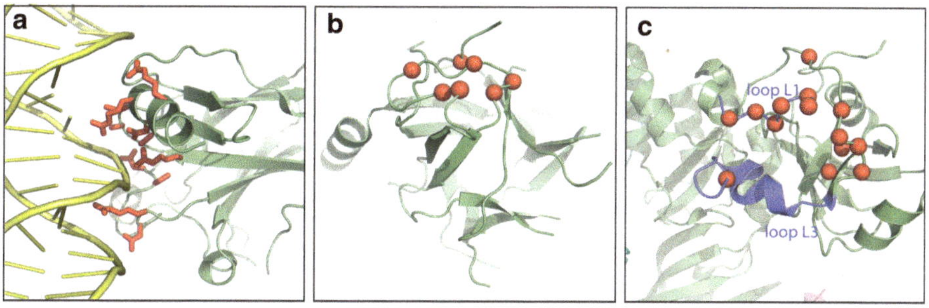

Fig. 4 Structure mapping of mutation hot spots. Mutation sites are shown in either *red sticks* or *spheres*. (**a**) p53 central domain in complex with DNA (PDB code 1TSR); (**b**) ryanodine receptor type 1 N-terminal domain (3HSM); and (**c**) yeast Cbf5 structure that is homologous to human dyskerin (3U28). The eukaryote-unique N-terminal extension (loops 1 and 3) is coloured *blue*

this article is accompanied with a web-based molecular viewer, allowing the reader to navigate the hot spot regions and each individual mutation along the protein landscape, in an interactive manner [62].

2.3.4 Lessons from Large-Scale Structural "Catalogues"

Taking advantage of the rapidly growing genomic and structural data, there are now efforts being made to catalogue missense variants on a large scale using vast protein datasets. Some of these efforts have focused on disease relevant protein families, such as kinases. For example, Lahiry et al. [63] used structural analysis of kinase mutations to correlate their locations on the protein to disease states. They observed that: (1) neutral mutations/polymorphisms, those that did not tend to cause disease, generally clustered in the C-terminal regions of the catalytic core, a region thought to have a basic structural role; (2) germline disease causing mutations, which cause metabolic disorders or loss-of-function developmental disorders, tended to cluster in the catalytic core in sites involved in regulation and substrate binding, as well as in protein–protein and allosteric interactions; (3) cancer causing somatic mutations were concentrated around the ATP binding and catalytic residues, directly influencing catalysis and resulting in the activation of oncogenes or deactivation of tumour suppressors.

Other studies have employed large datasets of missense variations spanning different protein families in order to detect any trends, consensus or "rules" which dictate whether certain types of amino acid changes will result in neutral polymorphisms or pathogenic mutations [64, 65]. These large-scale studies have generally arrived at the conclusion that pathogenic variants are more likely located in solvent-buried core regions, conserved residue positions, residues that contribute hydrogen bonds and those that alter more dramatically the physico-chemical properties of amino acids [64, 65]. Khan et al. [66] also looked at the distribution and frequency of pathogenic variations and found that arginine and glycine are the most mutated residue types, while overall mutability (i.e. the likelihood of being introduced in missense variations) is highest for cysteine and tryptophan. Using similar approaches, Hurst et al. [65] found that mutations of glycine, cysteine and tryptophan were more likely to be pathogenic than others, confirming results from previous small-scale studies. They also provided online access to their large database of structurally-mapped missense variations (www.bioinf.org.uk/saap/db). Taken together, these proteome-wide structure-based mutation analyses will continue to help us formulate better rules for our prediction of whether an uncharacterized amino acid variation will be pathogenic or not, thereby improving disease diagnostics in the future.

3 Protein Structure Analysis in Drug Development

3.1 Structural Biology and Target-Centric Drug Development

The opportunities presented from the post-genome era have also transformed rapidly the field of drug development. We are now made aware of the unprecedented number of potential therapeutic proteins, estimated in one study as reaching 10% of the predicted coding regions in the human genome (i.e. ~3,000 proteins) [67]. On the other hand, the current FDA-approved drugs target only a small number (~300) of human proteins or proteins from other pathogenic organisms [68–70]. This has made a fundamental impact in the direction of biomedical research in steering towards a more target-centric approach to bridge this gap. The main focus in this approach is to identify therapeutically-relevant drug targets that meet the double criteria of being disease-linked i.e. it has a causative role in the onset and/or progression of a disease, and being druggable i.e. it can be bound and modulated by a small molecule.

At the same time, the current field of drug development is facing tremendous challenges with ever-increasing research and development costs (reaching in some estimates up to $2 billion per drug [71]) and high attrition rate along the entire pipeline [72], where many potential projects fail through the early stages of hit identification and optimization to lead. As a result, the pharmaceutical industry is under continuous pressure to look for novel, high confidence disease targets and alternative drug design approaches. Amenable to the target-centric approach while having potentials in addressing some of the challenges in the pharmaceutical industry, the field of structural biology has been playing an increasing role in drug development, particularly at the early stages ("drug discovery"). Today, using structures to identify new lead compounds and as a basis for rational drug design is an integral part of many a drug development project.

3.2 Early Structural Applications in Lead Optimization

Before the technological advances in the past decade that have made protein structure determination faster and more cost-effective, the use of structure information in drug development in the 1980s and early 1990s has been confined to the lead optimization stage, in directing the chemical alterations of initial compound hits to improve their affinity, potency and selectivity. In this process, the protein structure of interest is determined in complexes with lead compounds identified from a high throughput screening (HTS) campaign, accomplished either by co-crystallizing the protein solution pre-incubated with the lead molecule, or by soaking pre-formed crystals of the apo protein with the ligand solution. The determined structure of the protein–ligand complex reveals the modes of interaction between the protein and ligand at the atomic level, e.g. short-range interactions such as hydrogen bonds, salt bridges, and hydrophobic contacts, the distances between the various interacting groups and atoms, and the presence of water molecules at the protein–ligand interaction site. This information is used to guide further iterative rounds of chemistry optimization and protein–ligand structure determination to establish a structure–activity relationship.

The first marketed drug developed via this structure-based approach was captopril, an inhibitor for angiotensin converting enzyme for the treatment of hypertension and congestive heart failure. This drug was designed in the mid-1970s on the basis of the homologous carboxypeptidase A protein which had been structurally characterized at the time [73]. Today, structure-based design approaches have delivered drugs to the market for a wide range of diseases, including retroviral [74, 75], glaucoma [76], influenza [77, 78] as well as cancer [79, 80] (Fig. 5). With advances, particularly in crystallography, the timeframe of protein structure determination is now sufficiently short to be amenable for many other stages of the drug discovery pipeline. As a result, the tools of structural analysis that were traditionally used in lead optimization are now being exploited to assist the processes of target identification, assessment of target druggability, and hit identification (Fig. 6), as outlined below.

3.3 Use of Structures in Target Identification

An early consideration in the target-centric approach of drug discovery is to identify and prioritize therapeutically-important proteins in the genome. Obtaining structural information at this stage, in the apo- or relevant liganded states of the protein, is an important milestone in target identification. High resolution atomic structures of many therapeutic targets are now available in the public domain, including kinases (e.g. AMPK [81]), viral proteins (influenza polymerase [82]), cytochrome P450 [83], metabolic enzymes (acetyl-CoA carboxylase [84]) and G-protein coupled receptors (β1-adrenergic receptor [85]). This unprecedented wealth of structure information helps establishing sequence–structure–function relationship and assessing potential ligand-binding capabilities, and is now part of the essential

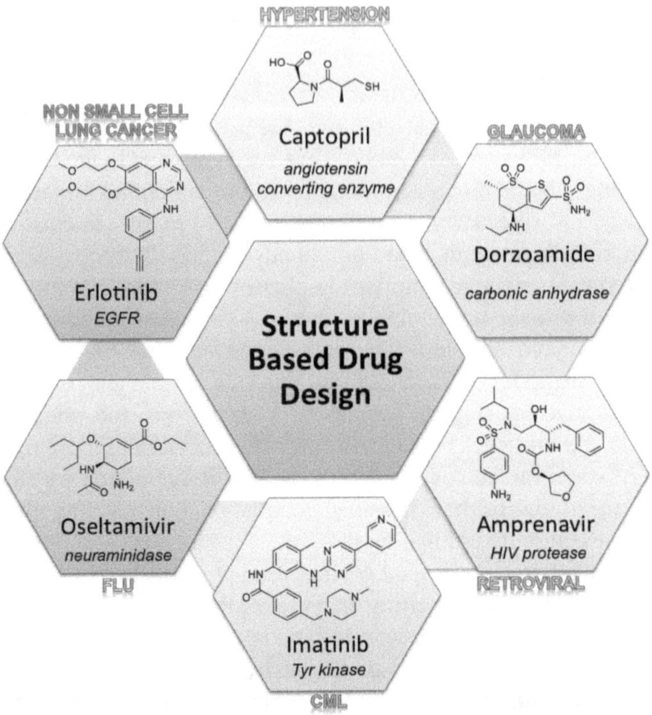

Fig. 5 Examples of structure-based drug design. FDA-approved drugs that have been derived from structure-based approaches. For each drug, its generic name, chemical structure, protein targeted and disease area applied is shown (references in the main text). *CML* chronic myeloid leukaemia; *EGFR* epidermal growth factor receptor

toolkit to complement in vivo target validation experiments (e.g. RNA interference screens, animal models, gene knockouts). The increase in available structures in the PDB also spurs the development of computational methods combining sequence and structural information to probe biological functions [86].

It is with the technological advances in structural biology that the field of "structural genomics" (SG) was born, to determine systematically 3D structures of proteins encoded in a genome primarily by crystallography and NMR. The overall objectives are to provide a structure coverage of the "protein universe" [87], to help define protein functions that cannot be predicted from sequences alone [88], and to facilitate the discovery, as well as selection, of genomic targets for drug therapy [89]. A number of large-scale SG efforts have emerged over the past 10 years, including RIKEN in Japan (www.riken.co.jp), Structure Proteomics in Europe (SPINE, www.spineurope.org), the Structural Genomics Consortium based in UK, Sweden and Canada (SGC, www.thesgc.com), as well as the Protein Structure Initiative in USA (PSI; www.nigms.nih.gov/psi). While sharing similar high-throughput methodologies and open access policy to their data, these SG initiatives differ in their scope and criteria for their target selection. A number of SG initiatives (e.g. PSI, RIKEN) aim to explore the novel protein folds that cannot

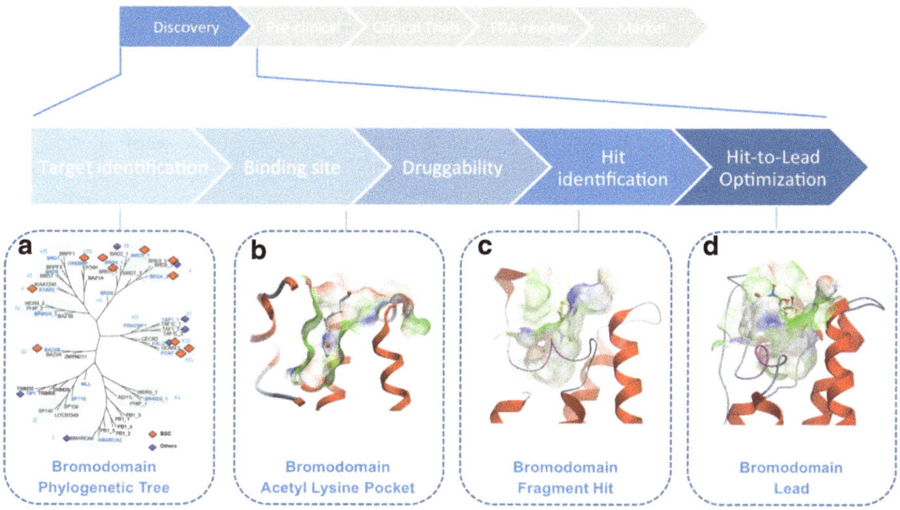

Fig. 6 Modern day structure-guided drug discovery. Protein structure analysis is nowadays incorporated into all early stages of drug development, including (**a**) target identification; (**b**) assessing binding site druggability; (**c**) hit identification, and (**d**) lead optimization. Example shown is from the chemical probe program at the Structural Genomics Consortium to develop small molecule binders for the family of histone-binding bromodomains (cf. [93] in main text)

be predicted from sequence [90], and subsequently leverage structure completeness of a genome by homology modeling of the remaining homologous proteins. Other SG programs take a biology-driven avenue, placing the emphasis more on medical relevance. For example, the Tuberculosis Structural Genomics Consortium [91] adopts an organism-based approach focusing on the obligate human pathogen *Mycobacterium tuberculosis*. To date nearly 10% of all proteins from the pathogen have been structurally characterized [92], which unravel a number of previously unannotated proteins as potential anti-tuberculosis targets. The human proteome-focused SGC studies protein families with therapeutic importance such as kinases, phosphatases and metabolic enzymes [93, 94]. These studies reveal structure-–function relationships between family members with regards to active site and substrate specificity (Fig. 6a), and emphasize their application to develop member/family-specific chemical probes and inhibitors [95].

3.4 *Use of Structures in Assessing Druggability*

3.4.1 Binding Site Detection

With the structural information of potential therapeutic targets made available, the next step in drug discovery is the identification of binding sites that are receptive to small molecule binding (Fig. 6b). The large repertoire of protein–ligand complexes in the PDB has provided a structural view of a ligand binding site to be a small pocket or invagination on the protein, accessible to the surface exterior, where ligands can fit to mediate a biological function. This pocket should harbour amino acid side-chains that contribute to hydrogen bonds and hydrophobic contacts. Based on these concepts, a number of pocket identification software have been developed to detect binding sites on the protein structure, adopting two general approaches (see [96] and references therein). The geometry-based methods (e.g. SURFNET, LIGSITE) look for geometrically-complex regions on the protein as natural binding sites tend to be concave surface invaginations. The probe/energy-based methods (e.g. GRID, AutoLigand, ICM) calculate the interaction energy between a probe molecule and protein at different point locations to define regions with favourable interaction energies.

A thorough understanding of the binding pocket space helps not only to assess its potential for drug binding but also to annotate functionally under-characterized proteins (i.e. de-orphanization). For example, delineating residues involved at the ligand binding site can stimulate site-directed mutagenesis experiments to probe their catalytic or regulatory roles. Structural characterization of binding sites also reveals the ligand-induced conformational changes on the protein target, which can range from small side-chain adjustment to whole-domain rearrangement [97]. Binding pockets are therefore not static sites as revealed in a structural snapshot, but dynamic regions important for the protein function. The conformational plasticity of the ligand binding sites needs to be addressed during structural analysis and drug design.

3.4.2 Druggability Index

The next step following pocket detection is an evaluation of whether it has the shape and chemical complementarity to accommodate high-affinity, drug-like molecules. This likelihood prediction of drug binding ("druggability") is crucial to target selection in drug discovery, with the hope of screening out unlikely candidates at an early stage. The emerging concept of protein druggability [98] is an extension to the "drug-likeness" rule-of-five for small molecules that attributes good oral bioavailability of drug compounds to certain favourable physico-chemical parameters [99]. Research groups are developing tools to predict druggability and quantify it in a "druggability index" using different structure-based metrics. Some correlate druggability with hit rates obtained from NMR screening of small fragments [100], whereas others base their predictions on binding affinity calculations [101] or on comparison of binding sites between different proteins/families that bind the same ligand to identify hot spot residues [102]. Druggability indices are especially useful in identifying non-native small molecule binding sites such as between protein–protein interaction surfaces [103].

3.5 Use of Structures in Hit Identification

A therapeutic protein that satisfies the criteria of disease linkage and druggability can enter the pipeline of a drug discovery program to identify hit compounds that bind the target and exert an effect. Traditionally this has been achieved by HTS [104]. In this approach a vast library collection of physically available compounds, accumulated by large pharmaceuticals over many years of research, isolated from natural sources or synthesized from combinatorial chemistry, is experimentally tested on the protein target using a high-density assay that measures either binding to or biochemical modulation of a protein. The aim is to identify compounds with IC_{50} values better than, e.g. 10 μM for further hit-to-lead optimization. The power of HTS relies on the implementation of a robust and sensitive assay and the interrogation of a vast compound collection, both requirements consuming significant resources in materials, time and manpower. Its success in generating hits also depends upon target classes, robustness of the assay and propensity to deliver false positives [105]. With these challenges under consideration, novel approaches continue to be explored as complement to the HTS method in hit discovery. In particular, *in silico* methods exploiting structural information of the binding pocket space are being widely explored. To this end, three structure-based approaches, namely virtual screening, de novo design and fragment-based screening, are gaining promise and are nowadays incorporated into almost every drug discovery project (Fig. 6c).

3.5.1 Virtual Screening

Virtual screening (VS) is often considered as the computational alternative to the classic HTS, hence its alias "virtual HTS" (see [106] and references therein). VS interrogates large chemical libraries *in silico*, often available as public compound databases, to predict their binding mode and affinity towards the protein structure. The prediction is based on docking calculations and generally involves two steps. First, every compound in the library is individually placed onto the protein pocket to generate different conformations and orientations ("poses") by sampling through the pocket space, taking into account ligand and protein flexibility at the pocket. Second, the binding modes between target and the ligand in its different poses are evaluated by a scoring function, and subsequently ranked to identify binding hits from the highest-scoring ligands and poses.

A rigorous scoring function is crucial to a VS campaign, so that it allows proper enrichment of true compound hits among the top ranking scores. Many scoring functions are developed, taking into account the interaction energies between ligand and protein ("force field-based") or statistical observations from experimentally derived protein–ligand structures with the basic premise that true hits share common protein–ligand interactions ("knowledge-based"). Nowadays a variety of docking software is available (e.g. DOCK, GOLD, AutoDock; see [106, 107] and references therein), each incorporated with different scoring functions. Current challenges in the docking tools include the need to improve scoring function accuracy, to take into account the various protonation, tautomerization and ionization states of compounds, and to predict ligand-induced protein conformations [107].

The strength of the structure-based VS approach is attributable to its capability to screen large databases (e.g. millions) of compounds with minimal computational power, more quickly and less expensively than HTS. Successful VS examples in hit identification include the development of EGFR inhibitors towards cancer cells [108], cysteine protease inhibitors of the SARS virus [109], and dihydroorotate dehydrogenase (DHODH) inhibitors towards rheumatoid arthritis [110]. The approaches of VS and HTS, with their mechanistic parallels, can also complement each other and have been applied side-by-side on the same drug development project [111] to facilitate hit identification.

3.5.2 De Novo Ligand Design

Structural knowledge of the binding pocket space can also guide the building of novel lead compounds from scratch [112, 113]. This de novo approach of drug design is not constrained by the known chemical structures from existing compounds, opening up the possibility of developing novel chemotypes [114]. The most common strategy is receptor/target-based de novo design, using a priori structural information of the target protein and its binding pocket. In this strategy, small building blocks (known as seeds or fragments) are positioned onto key interaction regions within the pocket, either by computational docking (as in Sect. 3.5.1), or recently by experimental methods such as crystallography and NMR (see Sect. 3.5.3). Each fragment can then be extended towards the neighbouring available space to build a lead compound that matches the binding pocket sterically and electrostatically ("growing" approach). Alternatively, multiple fragments bound independently at different but proximal regions of the pocket can be assembled into a lead compound using linker scaffolds ("linking" approach). A number of de novo drug design projects have yielded potential compound hits. For example, Heikkila et al. [115] exploited a species-specific hydrophobic ligand pocket on the *Plasmodium falciparum* DHODH protein to design potent parasite-specific compounds with an IC_{50} value of 43 μM. Ni et al. [116] developed inhibitors for the peptidylprolylisomerase cyclophilin A with IC_{50} values of 31.6 nM, with potentials as immunosuppressive agents. An important caveat of de novo design is that it often generates complex ligands with poor synthetic accessibility and pharmacokinetic properties. This is being addressed by software development to place emphasis on generating drug-like, synthetically-possible compounds.

3.5.3 Structure Based Fragment Screening

The X-ray and NMR methods of structure determination have also played a crucial role in a paradigm fragment-based screening approach [117]. Its premise is to screen experimentally hundreds to thousands of small compounds (usually between 100–300 Da in size) in order to identify low-affinity fragments (K_d in high μM range) that bind to different regions of the binding pocket, as a starting point for hit optimization. The subsequent optimization of fragment hits into a single hit compound can be rationalized, as in the de novo method, by the "growing" and "linking" processes. The concept of starting with small fragments is an appealing alternative to the conventional HTS attempts, with a number of merits. Fragments with their relatively small sizes and low complexity have been shown to provide higher hit rates than larger drug-like compounds from conventional screens [118], and can be optimized more efficiently. Fragments also allow a broader, more efficient sampling of the chemical space using a much smaller set of compounds (e.g. 100 fragments are equivalent to a 1,000,000 combinatory library) [114].

The relative weak binding of fragments (e.g. ~100 μM to 10 mM against target protein), which may be missed by a conventional HTS assay, can be experimentally determined by crystallography, NMR and other biophysical methods such as surface plasmon resonance [119, 120]. With inherently higher hit rates and the likelihood of multiple binding modes for a fragment hit, it is necessary to have its binding mode characterized from crystallography or NMR to allow hit-to-lead compound design. In particular, crystallography with its low-cost, high-throughput implementation is well attuned to fragment-based screening, allowing fast structure determination of protein-fragment complexes (Fig. 6d). A recent survey showed that 15 selective and potent inhibitors generated from fragment-based screening entered the phase I or II clinical trials. Examples include inhibitors for matrix metalloproteinase [121], aurora kinase [122], cyclin-dependent kinase 2 [123] and peroxisome proliferator-activated receptor [124]. An excellent update on fragment screening success examples across industry and academia was recently published [125].

4 Conclusion, Challenges and Future Perspectives

Over the past decade, the field of protein structural biology has responded to the challenging demands presented in the post-sequencing era by two revolutionary accomplishments. It has attained technological advances in the methods of structure determination in order to streamline the gene-to-structure process in a parallel, automated and miniaturized platform. Protein structures are now being solved by numerous academic and industrial research groups worldwide, on a daily or weekly basis. Structural biology has also broadened its scientific impact, successfully transforming itself from mere providers of structure information into an essential toolkit for molecular geneticists in the characterization and understanding of diseases, and for medicinal chemists to assist all stages of the drug discovery process. With its continuing scientific contribution and technical improvements, structural biology is more ready now than ever to offer promise in some of the biological areas that have so far proven difficult (Sects. 4.1 and 4.2), and to open up new exciting avenues for its applications (Sect. 4.3).

4.1 Studying Protein–Protein Interactions

A myriad of cellular processes are mediated by protein–protein interactions (e.g. in signaling, metabolism, cellular structure and transport), often requiring the formation of multiprotein macromolecular machineries. A mechanistic understanding of these biological processes therefore requires an examination of the protein complexes at the molecular level. Experimental methods such as X-ray crystallography, NMR and EM are now being used to complement biochemical and biophysical methods such as yeast two-hybrid, immuno-precipitation and fluorescence resonance energy transfer, to understand these interactions better. However, complex structure determination remains challenging as compared to its single protein counterpart, and often requires systematic mapping and delineation of the interacting region to obtain co-purified and co-crystallized complexes [126, 127]. This substantial investment in time and effort is reflected by the number of protein complex structures in the PDB being only one-sixth of single protein structures. In the absence of co-crystal structures, *in silico* methods serve as a promising alternative to generate complex structure models by protein–protein docking and homology modeling, and will continue to attract considerable attention and research due to their comparative ease of use [128]. The identification of druggable protein–protein interactions that participate in diseases also represents an exciting avenue in drug discovery. Targeting a protein–protein interface for small molecule modulation is often considered less tractable than conventional single protein targets, due to the large interacting surface and less pocket-like features. Nevertheless, over the years a number of protein–protein interaction inhibitors have been developed, assisted by available structural information of both protein–protein complexes and of individual proteins (e.g. interaction partners of interleukin IL-2, B-cell lymphoma 2 Bcl-X_L and human papilloma virus transcription factor E2; [129] and references therein), and some are now entering clinical trials.

4.2 The High Hanging-Fruits of Membrane Protein Structures

In addition to multiprotein complexes, many classes of disease-associated and therapeutically important proteins remain refractory to the current methods of structure determination. Particularly in mind are the integral membrane proteins, such as the family of G-protein coupled receptors (GPCR) that are predicted targets for ~30–50% of marketed drugs [68], and hence a major focus in pharmaceutical research. However, due to intrinsic difficulties with membrane protein crystallization, understanding GPCR structure and function has largely been achieved by homology modeling approaches. Recent structural breakthroughs, e.g. in the use of heterologous expression systems and in engineering mutations to stabilize proteins for crystallization [130], have brought the current number of available GPCR structures to six, an important increase, yet still in stark contrast to the total number of GPCRs predicted in the human genome (>900). Nevertheless, the structure determination over the past few years of a few highly-relevant GPCR drug targets (e.g. β1- and β2-adrenergic receptors [85, 131], A2A adenosine receptor [132], chemokine receptor CXCR4 [133] and dopamine D3 receptor [134]) has provided hope for structure-based methods to be applied routinely in GPCR drug discovery. These new structures offer promising opportunities for *in silico* compound screening and docking, and provide a diversity of available templates for homology modeling which, until the day that routine membrane protein crystallization has arrived, will continue to play a key role in leveraging structural coverage for this protein family.

4.3 Combining Mutation Analysis and Drug Design: Pharmacological Chaperones

An excellent example of combining the structural applications in mutation analysis and small molecule design is found in the emerging field of pharmacological chaperone therapy (PCT), a paradigm approach to treat inherited diseases that affect enzyme stability and function, such as phenylketonuria (cf. Sect. 2.3.1) and lysosomal storage disorders [135]. PCT involves the use of small molecules, often active site inhibitors or substrate mimics of the native protein, to stabilize mutant enzymes suffering from folding and trafficking defects. A great deal of ground work and proof-of-principle studies has incorporated structural information in order to establish the molecular basis of disease mutations and to identify those chaperone-responsive mutations with potential for PCT. To this end, a small-molecule screening effort to identify stabilizing therapeutic agents to treat PKU has already yielded two promising compounds for PAH stabilization [136]. Recently, crystal structures for a number of lysosomal hydrolases (e.g. β-hexosaminidase B [137] and acid β-glucosidase [138]) have been determined in complexes with pharmacological chaperones identified from chemical screening, to provide atomic insights into their modes of stabilization. The structure determination itself of these lysosomal enzymes is no small feat due to their heavily-glycosylated nature. The stage is now set for a systematic, structure-assisted approach in developing the next generation of chaperone compounds into clinical applications. The current work additionally reveals the potential of PCT as a general strategy to treat a wide range of rare genetic diseases, many of which are being unraveled by the year, and illustrates how structural biology has suitably positioned itself within the translational approach from bench to clinic.

Acknowledgements We thank Laura Spagnolo (University of Edinburgh) for the picture contribution in Fig. 3. The Structural Genomics Consortium is a registered charity (number 1097737) that receives funds from the Canadian Institutes for Health Research, the Canada Foundation for Innovation, Genome Canada, GlaxoSmithKline, Lilly Canada, the Novartis Research Foundation, Pfizer, Abbott, Takeda, the Ontario Ministry of Research and Innovation and the Wellcome Trust.

References

1. Kendrew JC, Bodo G, Dintzis HM, Parrish RG, Wyckoff H, Phillips DC (1958) A three-dimensional model of the myoglobin molecule obtained by X-ray analysis. Nature 181(4610):662–666
2. Savitsky P, Bray J, Cooper CD, Marsden BD, Mahajan P, Burgess-Brown NA et al (2010) High-throughput production of human proteins for crystallization: the SGC experience. J Struct Biol 172(1):3–13
3. Page R, Stevens RC (2004) Crystallization data mining in structural genomics: using positive and negative results to optimize protein crystallization screens. Methods 34(3):373–389
4. Joachimiak A (2009) High-throughput crystallography for structural genomics. Curr Opin Struct Biol 19(5):573–584
5. Manjasetty BA, Turnbull AP, Panjikar S, Bussow K, Chance MR (2008) Automated technologies and novel techniques to accelerate protein crystallography for structural genomics. Proteomics 8(4):612–625
6. Pellecchia M, Sem DS, Wuthrich K (2002) NMR in drug discovery. Nat Rev Drug Discov 1(3):211–219
7. Billeter M, Wagner G, Wuthrich K (2008) Solution NMR structure determination of proteins revisited. J Biomol NMR 42(3):155–158
8. Frank J (2002) Single-particle imaging of macromolecules by cryo-electron microscopy. Annu Rev Biophys Biomol Struct 31:303–319
9. Marti-Renom MA, Stuart AC, Fiser A, Sanchez R, Melo F, Sali A (2000) Comparative protein structure modeling of genes and genomes. Annu Rev Biophys Biomol Struct 29:291–325
10. Chothia C, Lesk AM (1986) The relation between the divergence of sequence and structure in proteins. EMBO J 5(4):823–826
11. Cavasotto CN, Phatak SS (2009) Homology modeling in drug discovery: current trends and applications. Drug Discov Today 14(13–14):676–683
12. Yue P, Li Z, Moult J (2005) Loss of protein structure stability as a major causative factor in monogenic disease. J Mol Biol 353(2):459–473
13. Yip YL, Scheib H, Diemand AV, Gattiker A, Famiglietti LM, Gasteiger E et al (2004) The Swiss-Prot variant page and the ModSNP database: a resource for sequence and structure information on human protein variants. Hum Mutat 23(5):464–470
14. Tramontano A, Morea V (2003) Assessment of homology-based predictions in CASP5. Proteins 53(Suppl 6):352–368
15. Metzker ML (2010) Sequencing technologies – the next generation. Nat Rev Genet 11(1):31–46
16. Ng SB, Nickerson DA, Bamshad MJ, Shendure J (2010) Massively parallel sequencing and rare disease. Hum Mol Genet 19(R2):R119–R124
17. Ku CS, Naidoo N, Pawitan Y (2011) Revisiting Mendelian disorders through exome sequencing. Hum Genet 129(4):351–370

18. Ng SB, Buckingham KJ, Lee C, Bigham AW, Tabor HK, Dent KM et al (2010) Exome sequencing identifies the cause of a mendelian disorder. Nat Genet 42(1):30–35
19. Ng SB, Turner EH, Robertson PD, Flygare SD, Bigham AW, Lee C et al (2009) Targeted capture and massively parallel sequencing of 12 human exomes. Nature 461(7261):272–276
20. Ng SB, Bigham AW, Buckingham KJ, Hannibal MC, McMillin MJ, Gildersleeve HI et al (2010) Exome sequencing identifies MLL2 mutations as a cause of Kabuki syndrome. Nat Genet 42(9):790–793
21. International HapMap Consortium (2003) The International HapMap Project. Nature 426 (6968):789–796
22. Hamosh A, Scott AF, Amberger JS, Bocchini CA, McKusick VA (2005) Online Mendelian Inheritance in Man (OMIM), a knowledgebase of human genes and genetic disorders. Nucleic Acids Res 33(Database issue):D514–D517
23. Stenson PD, Ball EV, Mort M, Phillips AD, Shiel JA, Thomas NS et al (2003) Human Gene Mutation Database (HGMD): 2003 update. Hum Mutat 21(6):577–581
24. Thusberg J, Vihinen M (2009) Pathogenic or not? And if so, then how? Studying the effects of missense mutations using bioinformatics methods. Hum Mutat 30(5):703–714
25. Jordan DM, Ramensky VE, Sunyaev SR (2010) Human allelic variation: perspective from protein function, structure, and evolution. Curr Opin Struct Biol 20(3):342–350
26. Karchin R (2009) Next generation tools for the annotation of human SNPs. Brief Bioinform 10(1):35–52
27. Sunyaev S, Ramensky V, Bork P (2000) Towards a structural basis of human non-synonymous single nucleotide polymorphisms. Trends Genet 16(5):198–200
28. Miller MP, Kumar S (2001) Understanding human disease mutations through the use of interspecific genetic variation. Hum Mol Genet 10(21):2319–2328
29. Hicks S, Wheeler DA, Plon SE, Kimmel M (2011) Prediction of missense mutation functionality depends on both the algorithm and sequence alignment employed. Hum Mutat 32(6):661–668
30. Calabrese R, Capriotti E, Fariselli P, Martelli PL, Casadio R (2009) Functional annotations improve the predictive score of human disease-related mutations in proteins. Hum Mutat 30(8):1237–1244
31. Yue P, Moult J (2006) Identification and analysis of deleterious human SNPs. J Mol Biol 356(5):1263–1274
32. Lobo PA, Van Petegem F (2009) Crystal structures of the N-terminal domains of cardiac and skeletal muscle ryanodine receptors: insights into disease mutations. Structure 17(11):1505–1514
33. Lew ED, Bae JH, Rohmann E, Wollnik B, Schlessinger J (2007) Structural basis for reduced FGFR2 activity in LADD syndrome: implications for FGFR autoinhibition and activation. Proc Natl Acad Sci USA 104(50):19802–19807
34. Chaikuad A, Froese DS, Berridge G, von Delft F, Oppermann U, Yue WW (2011) Conformational plasticity of glycogenin and its maltosaccharide substrate during glycogen biogenesis. Proc Natl Acad Sci USA 108(52):21028–21033
35. Malay AD, Procious SL, Tolan DR (2002) The temperature dependence of activity and structure for the most prevalent mutant aldolase B associated with hereditary fructose intolerance. Arch Biochem Biophys 408(2):295–304
36. Zhang Z, Norris J, Schwartz C, Alexov E (2011) In silico and in vitro investigations of the mutability of disease-causing missense mutation sites in spermine synthase. PLoS One 6(5): e20373
37. Anderson PC, Daggett V (2008) Molecular basis for the structural instability of human DJ-1 induced by the L166P mutation associated with Parkinson's disease. Biochemistry 47 (36):9380–9393
38. Niesen FH, Berglund H, Vedadi M (2007) The use of differential scanning fluorimetry to detect ligand interactions that promote protein stability. Nat Protoc 2(9):2212–2221
39. Froese DS, Kochan G, Muniz JR, Wu X, Gileadi C, Ugochukwu E et al (2010) Structures of the human GTPase MMAA and vitamin B12-dependent methylmalonyl-CoA mutase and insight into their complex formation. J Biol Chem 285(49):38204–38213

40. Wu L, Pan L, Wei Z, Zhang M (2011) Structure of MyTH4-FERM domains in myosin VIIa tail bound to cargo. Science 331(6018):757–760
41. Bridwell-Rabb J, Winn AM, Barondeau DP (2011) Structure-function analysis of Friedreich's ataxia mutants reveals determinants of frataxin binding and activation of the Fe-S assembly complex. Biochemistry 50(33):7265–7274
42. Wang Z, Moult J (2001) SNPs, protein structure, and disease. Hum Mutat 17(4):263–270
43. Gregersen N, Bross P, Vang S, Christensen JH (2006) Protein misfolding and human disease. Annu Rev Genomics Hum Genet 7:103–124
44. Mitchell JJ, Trakadis YJ, Scriver CR (2011) Phenylalanine hydroxylase deficiency. Genet Med 13(8):697–707
45. Dobson CM (2004) Principles of protein folding, misfolding and aggregation. Semin Cell Dev Biol 15(1):3–16
46. Jennings IG, Cotton RG, Kobe B (2000) Structural interpretation of mutations in phenylalanine hydroxylase protein aids in identifying genotype-phenotype correlations in phenylketonuria. Eur J Hum Genet 8(9):683–696
47. Erlandsen H, Stevens RC (1999) The structural basis of phenylketonuria. Mol Genet Metab 68(2):103–125
48. Dobrowolski SF, Pey AL, Koch R, Levy H, Ellingson CC, Naylor EW et al (2009) Biochemical characterization of mutant phenylalanine hydroxylase enzymes and correlation with clinical presentation in hyperphenylalaninaemic patients. J Inherit Metab Dis 32(1):10–21
49. Pey AL, Stricher F, Serrano L, Martinez A (2007) Predicted effects of missense mutations on native-state stability account for phenotypic outcome in phenylketonuria, a paradigm of misfolding diseases. Am J Hum Genet 81(5):1006–1024
50. Gersting SW, Kemter KF, Staudigl M, Messing DD, Danecka MK, Lagler FB et al (2008) Loss of function in phenylketonuria is caused by impaired molecular motions and conformational instability. Am J Hum Genet 83(1):5–17
51. Matthews BW (1993) Structural and genetic analysis of protein stability. Annu Rev Biochem 62:139–160
52. Xiao J, Madhan B, Li Y, Brodsky B, Baum J (2011) Osteogenesis imperfecta model peptides: incorporation of residues replacing Gly within a triple helix achieved by renucleation and local flexibility. Biophys J 101(2):449–458
53. Tang NL, Hui J, Young E, Worthington V, To KF, Cheung KL et al (2003) A novel mutation (G233D) in the glycogen phosphorylase gene in a patient with hepatic glycogen storage disease and residual enzyme activity. Mol Genet Metab 79(2):142–145
54. Wu WW, Molday RS (2003) Defective discoidin domain structure, subunit assembly, and endoplasmic reticulum processing of retinoschisin are primary mechanisms responsible for X-linked retinoschisis. J Biol Chem 278(30):28139–28146
55. Malay AD, Allen KN, Tolan DR (2005) Structure of the thermolabile mutant aldolase B, A149P: molecular basis of hereditary fructose intolerance. J Mol Biol 347(1):135–144
56. Jeyabalan J, Nesbit MA, Galvanovskis J, Callaghan R, Rorsman P, Thakker RV (2010) SEDLIN forms homodimers: characterisation of SEDLIN mutations and their interactions with transcription factors MBP1, PITX1 and SF1. PLoS One 5(5):e10646
57. Joerger AC, Fersht AR (2007) Structural biology of the tumor suppressor p53 and cancer-associated mutants. Adv Cancer Res 97:1–23
58. Cho Y, Gorina S, Jeffrey PD, Pavletich NP (1994) Crystal structure of a p53 tumor suppressor-DNA complex: understanding tumorigenic mutations. Science 265(5170):346–355
59. Amador FJ, Liu S, Ishiyama N, Plevin MJ, Wilson A, MacLennan DH et al (2009) Crystal structure of type I ryanodine receptor amino-terminal beta-trefoil domain reveals a disease-associated mutation "hot spot" loop. Proc Natl Acad Sci USA 106(27):11040–11044
60. Li S, Duan J, Li D, Yang B, Dong M, Ye K (2011) Reconstitution and structural analysis of the yeast box H/ACA RNA-guided pseudouridine synthase. Genes Dev 25(22):2409–2421
61. Picaud S, Kavanagh KL, Yue WW, Lee WH, Muller-Knapp S, Gileadi O et al (2011) Structural basis of fumarate hydratase deficiency. J Inherit Metab Dis 34(3):671–676

62. Lee WH, Yue WW, Raush E, Totrov M, Abagyan R, Oppermann U et al (2011) Interactive JIMD articles using the iSee concept: turning a new page on structural biology data. J Inherit Metab Dis 34(3):565–567
63. Lahiry P, Torkamani A, Schork NJ, Hegele RA (2010) Kinase mutations in human disease: interpreting genotype-phenotype relationships. Nat Rev Genet 11(1):60–74
64. Gong S, Blundell TL (2010) Structural and functional restraints on the occurrence of single amino acid variations in human proteins. PLoS One 5(2):e9186
65. Hurst JM, McMillan LE, Porter CT, Allen J, Fakorede A, Martin AC (2009) The SAAPdb web resource: a large-scale structural analysis of mutant proteins. Hum Mutat 30(4):616–624
66. Khan S, Vihinen M (2007) Spectrum of disease-causing mutations in protein secondary structures. BMC Struct Biol 7:56
67. Hopkins AL, Groom CR (2002) The druggable genome. Nat Rev Drug Discov 1(9):727–730
68. Overington JP, Al-Lazikani B, Hopkins AL (2006) How many drug targets are there? Nat Rev Drug Discov 5(12):993–996
69. Imming P, Sinning C, Meyer A (2006) Drugs, their targets and the nature and number of drug targets. Nat Rev Drug Discov 5(10):821–834
70. Rask-Andersen M, Almen MS, Schioth HB (2011) Trends in the exploitation of novel drug targets. Nat Rev Drug Discov 10(8):579–590
71. Morgan S, Grootendorst P, Lexchin J, Cunningham C, Greyson D (2011) The cost of drug development: a systematic review. Health Policy 100(1):4–17
72. Paul SM, Mytelka DS, Dunwiddie CT, Persinger CC, Munos BH, Lindborg SR et al (2010) How to improve R&D productivity: the pharmaceutical industry's grand challenge. Nat Rev Drug Discov 9(3):203–214. doi:10.1038/nrd3078
73. Hassall CH, Krohn A, Moody CJ, Thomas WA (1982) The design of a new group of angiotensin-converting enzyme inhibitors. FEBS Lett 147(2):175–179
74. Lapatto R, Blundell T, Hemmings A, Overington J, Wilderspin A, Wood S et al (1989) X-Ray analysis of HIV-1 proteinase at 2.7 A resolution confirms structural homology among retroviral enzymes. Nature 342(6247):299–302
75. Miller M, Schneider J, Sathyanarayana BK, Toth MV, Marshall GR, Clawson L et al (1989) Structure of complex of synthetic HIV-1 protease with a substrate-based inhibitor at 2.3 A resolution. Science 246(4934):1149–1152
76. Supuran CT, Scozzafava A, Casini A (2003) Carbonic anhydrase inhibitors. Med Res Rev 23 (2):146–189
77. von Itzstein M, Wu W-Y, Kok GB, Pegg MS, Dyason JC, Jin B et al (1993) Rational design of potent sialidase-based inhibitors of influenza virus replication. Nature 363(6428):418–423. doi:10.1038/363418a0
78. Lew W, Chen X, Kim CU (2000) Discovery and development of GS 4104 (oseltamivir) an orally active influenza neuraminidase inhibitor. Curr Med Chem 7(6):663–672
79. Tokarski JS, Newitt JA, Chang CY, Cheng JD, Wittekind M, Kiefer SE et al (2006) The structure of Dasatinib (BMS-354825) bound to activated ABL kinase domain elucidates its inhibitory activity against imatinib-resistant ABL mutants. Cancer Res 66(11):5790–5797
80. Schindler T, Bornmann W, Pellicena P, Miller WT, Clarkson B, Kuriyan J (2000) Structural mechanism for STI-571 inhibition of abelson tyrosine kinase. Science 289(5486):1938–1942
81. Chen L, Jiao ZH, Zheng LS, Zhang YY, Xie ST, Wang ZX et al (2009) Structural insight into the autoinhibition mechanism of AMP-activated protein kinase. Nature 459(7250):1146–1149
82. Yuan P, Bartlam M, Lou Z, Chen S, Zhou J, He X et al (2009) Crystal structure of an avian influenza polymerase PA(N) reveals an endonuclease active site. Nature 458(7240):909–913
83. Williams PA, Cosme J, Vinkovic DM, Ward A, Angove HC, Day PJ et al (2004) Crystal structures of human cytochrome P450 3A4 bound to metyrapone and progesterone. Science 305(5684):683–686
84. Zhang H, Tweel B, Li J, Tong L (2004) Crystal structure of the carboxyltransferase domain of acetyl-coenzyme A carboxylase in complex with CP-640186. Structure 12(9):1683–1691
85. Warne T, Serrano-Vega MJ, Baker JG, Moukhametzianov R, Edwards PC, Henderson R et al (2008) Structure of a beta1-adrenergic G-protein-coupled receptor. Nature 454 (7203):486–491

86. Cammer SA, Hoffman BT, Speir JA, Canady MA, Nelson MR, Knutson S et al (2003) Structure-based active site profiles for genome analysis and functional family subclassification. J Mol Biol 334(3):387–401
87. Grabowski M, Chruszcz M, Zimmerman MD, Kirillova O, Minor W (2009) Benefits of structural genomics for drug discovery research. Infect Disord Drug Targets 9(5):459–474
88. Shin DH, Hou J, Chandonia JM, Das D, Choi IG, Kim R et al (2007) Structure-based inference of molecular functions of proteins of unknown function from Berkeley Structural Genomics Center. J Struct Funct Genomics 8(2–3):99–105
89. Weigelt J (2010) Structural genomics-impact on biomedicine and drug discovery. Exp Cell Res 316(8):1332–1338
90. Dessailly BH, Nair R, Jaroszewski L, Fajardo JE, Kouranov A, Lee D et al (2009) PSI-2: structural genomics to cover protein domain family space. Structure 17(6):869–881
91. Chim N, Habel JE, Johnston JM, Krieger I, Miallau L, Sankaranarayanan R et al (2011) The TB structural genomics consortium: a decade of progress. Tuberculosis (Edinb) 91 (2):155–172
92. Ehebauer MT, Wilmanns M (2011) The progress made in determining the *Mycobacterium tuberculosis* structural proteome. Proteomics 11(15):3128–3133
93. Yue WW, Oppermann U (2011) High-throughput structural biology of metabolic enzymes and its impact on human diseases. J Inherit Metab Dis 34(3):575–581
94. Edwards A (2009) Large-scale structural biology of the human proteome. Annu Rev Biochem 78:541–568
95. Filippakopoulos P, Qi J, Picaud S, Shen Y, Smith WB, Fedorov O et al (2010) Selective inhibition of BET bromodomains. Nature 468(7327):1067–1073
96. Perot S, Sperandio O, Miteva MA, Camproux AC, Villoutreix BO (2010) Druggable pockets and binding site centric chemical space: a paradigm shift in drug discovery. Drug Discov Today 15(15–16):656–667
97. Hammes-Schiffer S, Benkovic SJ (2006) Relating protein motion to catalysis. Annu Rev Biochem 75:519–541
98. Keller TH, Pichota A, Yin Z (2006) A practical view of 'druggability'. Curr Opin Chem Biol 10(4):357–361
99. Lipinski CA, Lombardo F, Dominy BW, Feeney PJ (2001) Experimental and computational approaches to estimate solubility and permeability in drug discovery and development settings. Adv Drug Deliv Rev 46(1–3):3–26
100. Hajduk PJ, Huth JR, Fesik SW (2005) Druggability indices for protein targets derived from NMR-based screening data. J Med Chem 48(7):2518–2525
101. Cheng AC, Coleman RG, Smyth KT, Cao Q, Soulard P, Caffrey DR et al (2007) Structure-based maximal affinity model predicts small-molecule druggability. Nat Biotechnol 25(1):71–75
102. Ciulli A, Williams G, Smith AG, Blundell TL, Abell C (2006) Probing hot spots at protein-ligand binding sites: a fragment-based approach using biophysical methods. J Med Chem 49(16):4992–5000
103. Wells JA, McClendon CL (2007) Reaching for high-hanging fruit in drug discovery at protein-protein interfaces. Nature 450(7172):1001–1009. doi:10.1038/nature06526
104. Bleicher KH, Bohm HJ, Muller K, Alanine AI (2003) Hit and lead generation: beyond high-throughput screening. Nat Rev Drug Discov 2(5):369–378
105. Macarron R, Banks MN, Bojanic D, Burns DJ, Cirovic DA, Garyantes T et al (2011) Impact of high-throughput screening in biomedical research. Nat Rev Drug Discov 10(3):188–195. doi:10.1038/nrd3368
106. Klebe G (2006) Virtual ligand screening: strategies, perspectives and limitations. Drug Discov Today 11(13–14):580–594
107. Kalyaanamoorthy S, Chen YP (2011) Structure-based drug design to augment hit discovery. Drug Discov Today 16(17–18):831–839

108. Cavasotto CN, Ortiz MA, Abagyan RA, Piedrafita FJ (2006) In silico identification of novel EGFR inhibitors with antiproliferative activity against cancer cells. Bioorg Med Chem Lett 16(7):1969–1974
109. Dooley AJ, Shindo N, Taggart B, Park JG, Pang YP (2006) From genome to drug lead: identification of a small-molecule inhibitor of the SARS virus. Bioorg Med Chem Lett 16 (4):830–833
110. McLean LR, Zhang Y, Degnen W, Peppard J, Cabel D, Zou C et al (2010) Discovery of novel inhibitors for DHODH via virtual screening and X-ray crystallographic structures. Bioorg Med Chem Lett 20(6):1981–1984
111. Ferreira RS, Simeonov A, Jadhav A, Eidam O, Mott BT, Keiser MJ et al (2010) Complementarity between a docking and a high-throughput screen in discovering new cruzain inhibitors. J Med Chem 53(13):4891–4905
112. Schneider G, Hartenfeller M, Reutlinger M, Tanrikulu Y, Proschak E, Schneider P (2009) Voyages to the (un)known: adaptive design of bioactive compounds. Trends Biotechnol 27(1):18–26
113. Hartenfeller M, Schneider G (2011) De novo drug design. Methods Mol Biol 672:299–323
114. Fink T, Bruggesser H, Reymond JL (2005) Virtual exploration of the small-molecule chemical universe below 160 Daltons. Angew Chem Int Ed Engl 44(10):1504–1508
115. Heikkila T, Thirumalairajan S, Davies M, Parsons MR, McConkey AG, Fishwick CW et al (2006) The first de novo designed inhibitors of *Plasmodium falciparum* dihydroorotate dehydrogenase. Bioorg Med Chem Lett 16(1):88–92
116. Ni S, Yuan Y, Huang J, Mao X, Lv M, Zhu J et al (2009) Discovering potent small molecule inhibitors of cyclophilin A using de novo drug design approach. J Med Chem 52(17):5295–5298
117. Blundell TL, Jhoti H, Abell C (2002) High-throughput crystallography for lead discovery in drug design. Nat Rev Drug Discov 1(1):45–54
118. Hann MM, Leach AR, Harper G (2001) Molecular complexity and its impact on the probability of finding leads for drug discovery. J Chem Inf Comput Sci 41(3):856–864
119. Hartshorn MJ, Murray CW, Cleasby A, Frederickson M, Tickle IJ, Jhoti H (2005) Fragment-based lead discovery using X-ray crystallography. J Med Chem 48(2):403–413
120. Shuker SB, Hajduk PJ, Meadows RP, Fesik SW (1996) Discovering high-affinity ligands for proteins: SAR by NMR. Science 274(5292):1531–1534
121. Wada CK, Holms JH, Curtin ML, Dai Y, Florjancic AS, Garland RB et al (2002) Phenoxyphenyl sulfone N-formylhydroxylamines (retrohydroxamates) as potent, selective, orally bioavailable matrix metalloproteinase inhibitors. J Med Chem 45(1):219–232
122. Howard S, Berdini V, Boulstridge JA, Carr MG, Cross DM, Curry J et al (2009) Fragment-based discovery of the pyrazol-4-yl urea (AT9283), a multitargeted kinase inhibitor with potent aurora kinase activity. J Med Chem 52(2):379–388
123. Wyatt PG, Woodhead AJ, Berdini V, Boulstridge JA, Carr MG, Cross DM et al (2008) Identification of N-(4-piperidinyl)-4-(2,6-dichlorobenzoylamino)-1H-pyrazole-3-carboxamide (AT7519), a novel cyclin dependent kinase inhibitor using fragment-based X-ray crystallography and structure based drug design. J Med Chem 51(16):4986–4999
124. Artis DR, Lin JJ, Zhang C, Wang W, Mehra U, Perreault M et al (2009) Scaffold-based discovery of indeglitazar, a PPAR pan-active anti-diabetic agent. Proc Natl Acad Sci USA 106(1):262–267
125. de Kloe GE, Bailey D, Leurs R, de Esch IJ (2009) Transforming fragments into candidates: small becomes big in medicinal chemistry. Drug Discov Today 14(13–14):630–646
126. Strong M, Sawaya MR, Wang S, Phillips M, Cascio D, Eisenberg D (2006) Toward the structural genomics of complexes: crystal structure of a PE/PPE protein complex from *Mycobacterium tuberculosis*. Proc Natl Acad Sci USA 103(21):8060–8065
127. Brooun A, Foster SA, Chrencik JE, Chien EY, Kolatkar AR, Streiff M et al (2007) Remedial strategies in structural proteomics: expression, purification, and crystallization of the Vav1/Rac1 complex. Protein Expr Purif 53(1):51–62

128. Mukherjee S, Zhang Y (2011) Protein-protein complex structure predictions by multimeric threading and template recombination. Structure 19(7):955–966
129. Wells JA, McClendon CL (2007) Reaching for high-hanging fruit in drug discovery at protein-protein interfaces. Nature 450(7172):1001–1009
130. Rosenbaum DM, Cherezov V, Hanson MA, Rasmussen SG, Thian FS, Kobilka TS et al (2007) GPCR engineering yields high-resolution structural insights into beta2-adrenergic receptor function. Science 318(5854):1266–1273
131. Cherezov V, Rosenbaum DM, Hanson MA, Rasmussen SG, Thian FS, Kobilka TS et al (2007) High-resolution crystal structure of an engineered human beta2-adrenergic G protein-coupled receptor. Science 318(5854):1258–1265
132. Jaakola VP, Griffith MT, Hanson MA, Cherezov V, Chien EY, Lane JR et al (2008) The 2.6 angstrom crystal structure of a human A2A adenosine receptor bound to an antagonist. Science 322(5905):1211–1217
133. Wu B, Chien EY, Mol CD, Fenalti G, Liu W, Katritch V et al (2010) Structures of the CXCR4 chemokine GPCR with small-molecule and cyclic peptide antagonists. Science 330 (6007):1066–1071
134. Chien EY, Liu W, Zhao Q, Katritch V, Han GW, Hanson MA et al (2010) Structure of the human dopamine D3 receptor in complex with a D2/D3 selective antagonist. Science 330 (6007):1091–1095
135. Chaudhuri TK, Paul S (2006) Protein-misfolding diseases and chaperone-based therapeutic approaches. FEBS J 273(7):1331–1349
136. Pey AL, Ying M, Cremades N, Velazquez-Campoy A, Scherer T, Thony B et al (2008) Identification of pharmacological chaperones as potential therapeutic agents to treat phenylketonuria. J Clin Invest 118(8):2858–2867
137. Bateman KS, Cherney MM, Mahuran DJ, Tropak M, James MN (2011) Crystal structure of beta-hexosaminidase B in complex with pyrimethamine, a potential pharmacological chaperone. J Med Chem 54(5):1421–1429
138. Lieberman RL, Wustman BA, Huertas P, Powe AC Jr, Pine CW, Khanna R et al (2007) Structure of acid beta-glucosidase with pharmacological chaperone provides insight into Gaucher disease. Nat Chem Biol 3(2):101–107

Top Curr Chem (2014) 336: 99–116
DOI: 10.1007/128_2012_327

Published online: 20 May 2012

Emerging Applications of Single-Cell Diagnostics

M. Shirai, T. Taniguchi, and H. Kambara

Abstract The performance of DNA sequencers (next generation sequencing) is rapidly enhanced these days, being used for genetic diagnostics. Although many phenomena could be elucidated with such massive genome data, it is still a big challenge to obtain comprehensive understanding of diseases and the relevant biology at the cellular level. In general terms, the data obtained to date are averages of ensembles of cells, but it is not certain whether the same features are the same inside an individual cell. Accordingly, important information may be masked by the averaging process. As the technologies for analyzing bio-molecular components in single cells are being developed, single cell analysis seems promising to address the current limitations due to averaging problems. Although the technologies for single cell analysis are still at the infant stage, the single cell approach has the potential to improve the accuracy of diagnosis based on knowledge of intra- and inter-cellular networks. In this review several technologies and applications (especially medical applications) of genome and transcriptome analysis or single cells are described.

Keywords Digital counting of cDNA · Digital PCR · Droplet microfluidics · Gene expression profile · Genome · Heterogeneity · Microfluidics · Next generation DNA sequencer (NGS) · PCR bias · Single cell analysis · Single cell cDNA library · Transcriptome

---

M. Shirai, T. Taniguchi, and H. Kambara (✉)
Central Research Laboratory, Hitachi, Ltd., 1-280, Higachi-koigakubo Kokubunji-shi, Tokyo, Japan
e-mail: hideki.kambara.se@hitachi.com

Contents

1 Introduction 100
2 Technologies for Analyzing Gene Expressions in Single Cells 101
2.1 Global Amplification of mRNA Coupled with DNA Microarray Analysis [72, 73] 102
2.2 Reusable Single-Cell cDNA Library on Beads Coupled with qPCR [16] 103
2.3 Multiple Gene Expression Analysis with a Microfluidic Device [27, 51, 73, 78–82] 106
2.4 Digital Counting of cDNA Molecules with Next-Generation Sequencings 107
3 Potential Applications of Single-Cell Analysis in the Clinic 108
References 111

1 Introduction

A major change in bio-medical fields has begun following the completion of the human genome project. The use of genome data for medical applications has been greatly promoted by the NIH "$1000 genome project," which encourages development of ultrahigh-performance DNA sequencers [1–5]. As a result of this project, many genomes, as well as transcriptome data, have been obtained [6–12] and used to elucidate various biological and medical phenomena. At the initial stage of the human genome project, it was a common belief that almost everything regarding human diseases could be elucidated if whole genome sequences were obtained. Although many phenomena could be elucidated with such massive genome data, many more new biological questions arise. As living organisms are not simply a collection of genetic codes, it is important to understand it as an interesting system on the basis of molecular science. There is a particular problem associated with most cellular measurements or diagnostic used nowadays, which is that the measurement only represents a mean reading obtained from a population of a fairly large number of cells (often more than hundreds or thousands). Therefore, the variation or range of measurements produced in any single cell is not obtained. It will not be an issue if every cell behaves in the same way. However, increasing evidence has shown that this is not the case and that there is a great deal of cellular heterogeneity in the example of primary tissue and cancer. Accordingly, important information may be lost during the averaging processes [13–17]. Therefore research on analyzing single cells is becoming important.

For example, it has been clarified that gene expressions of individual cells in a cell pool are not uniform even under uniform culture conditions [18]. The same situation has been reported for cancer tissue [19–22], ES cells [23–25], and hematopoietic stem cells [14, 26, 27]. For medical applications of single-cell analysis it is necessary to analyze many single cells and to figure out their interactions. As the technologies for analyzing bio-molecular components in single cells are still at an early stage of development, applications of single-cell analysis in medicine are largely exploratory research.

Data on individual cells may be observed via microscopy and molecular-probe methods using fluorescent probes [13, 19, 25, 28–34]. However, as the number of target molecules is low in a single cell, the number of analyzable targets that could be detected and quantitated by imaging technology is very limited. Therefore, the combination of imaging and whole-genome and transcriptome analysis with additional tools such as microfluidics and next-generation DNA sequencers may be more promising. Several technological approaches and applications of genome and transcriptome analysis developed for single cells are described here.

2 Technologies for Analyzing Gene Expressions in Single Cells

The targets of bio-molecular component analysis for single cells include genomic DNA, transcriptomes (RNA), proteomes (peptides and proteins), and metabolomes (metabolites). For the analysis of metabolites in single cells, mass spectrometry is used [26, 35–38]. The limiting issue concerning mass spectrometry is how to increase the detection sensitivity sufficiently for analyzing many metabolites at low quantities in single cells [26, 35–38]. For proteome analysis, single-molecule imaging technology is a powerful tool [39–43]. Recently, an imaging technology coupled with rolling-circle DNA amplification for observing single protein molecules in a cell has been reported [34].

For conventional gene expression analysis, DNA microarrays have mainly been used [44–49]. As it requires a large amount of mRNA or cDNA for the analysis, RNA or cDNA obtained from a single cell must be amplified. Amplification bias is common in PCR; however, Kurimoto et al. optimized the PCR conditions to minimize PCR bias for gene expression analysis of single cells [45]. They succeeded in reducing the PCR bias down to the range of three- to fourfold. On the other hand, the most accurate quantitative-analysis method is qPCR. However, qPCR requires the division of the RNA templates and qPCR solution into many fractions according to the number of target gene species, and such division is not feasible for analyzing genes expressed at single cell level [44]. Accordingly, we previously developed a reusable single-cell cDNA library on beads that can be used repeatedly for multiple gene expression analyses [16]. The combination of qPCR and linear cDNA amplification followed by division of the amplified product is used to make multiple gene expression analysis possible (commercialized by Fluidigm Corporation) [20, 50, 51]. Large-scale DNA sequencing with next-generation DNA sequencers has been applied for gene expression analyses of pooled-cell samples [8–12, 52–65]. Recently, several attempts to use next-generation DNA sequencers for digital counting of mRNA in a single cell have been reported [66–70]. Since MPS requires a large number of templates for sequencing, cDNA obtained from a single cell must be amplified prior to sequencing. To eliminate errors due to amplification bias, cell tags and molecular tags to identify cDNA molecules have been proposed as quality control markers [71]. These methods are outlined in the following sections.

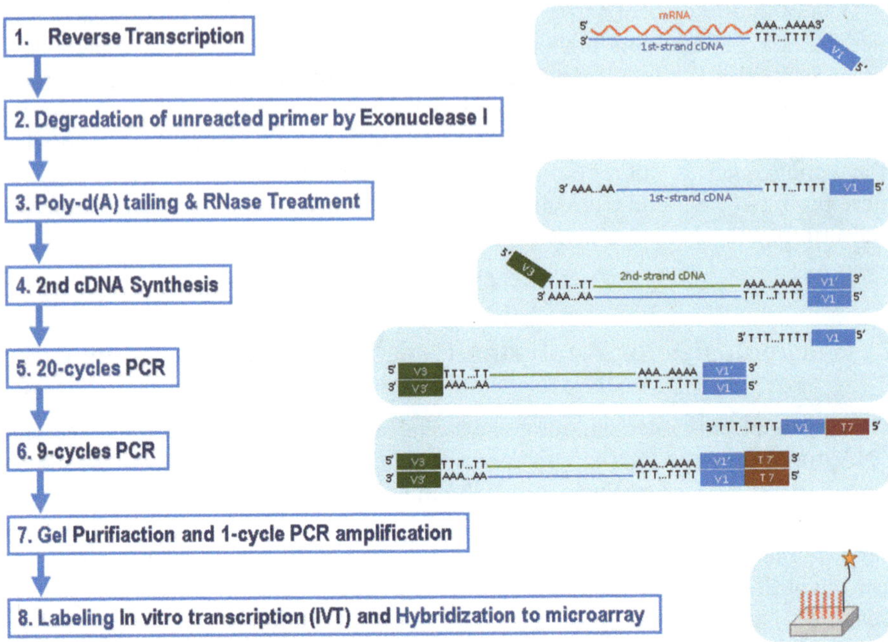

Fig. 1 Global cDNA amplification with reduced bias for downstream analysis with DNA microarray

2.1 *Global Amplification of mRNA Coupled with DNA Microarray Analysis [72, 73]*

The first trial of gene expression analysis for single cells by using DNA microarrays was done by Brady et al. However, they observed a large PCR amplification bias and therefore that analysis was not really quantitative. Kurimoto et al. improved the method by employing spike RNA, removing residual capture probes with Exonuclease I, and using low-cycle PCR (which gave a small amplification bias). They succeeded in reducing the PCR bias by a factor of less than 4 [45]. The method is outlined in Fig. 1.

Anchored poly-T primers are used for capturing mRNA to produce the first cDNA strands by reverse transcription. The anchored sequence V1 is used as the priming site for later PCR amplification. The excess primers are then digested by Exonuclease I to prevent the production of primer dimmers which disturb the amplification of the targets. The 3′ termini of the first cDNA are modified by poly-A tailing with terminal deoxynucleotidyl transferase (TdT) followed by PCR amplification. As the PCR process is very sensitive to the conditions and frequently causes amplification bias, the modified first cDNA strands are divided into four fractions as a quality control procedure for comparing errors in the following sequence of processes. The second anchored poly T primers are added to each fraction to hybridize on poly-A tails of cDNA for the second cDNA synthesis. The anchored sequence V3 is used as the priming site for the forward primers in the following PCR. A 20-cycle PCR is carried out with V3 (forward) and V1 (reverse) primers. A low number of PCR cycles are used for reducing amplification bias. An additional nine-cycle-PCR is then carried out with V3 primers, the primers having a V1 sequence as well as a T7-promoter sequence. Then the promoter sequence is used for in vitro transcription (IVT) which could yield ~5 μg of cRNA (i.e., enough for DNA microarray analysis).

The accuracy of the method, namely, cDNA amplification coupled with DNA microarray, was checked by sequencing and by qPCR for 40 genes. It was confirmed that the product sizes were in the range 388–1,652 and that the amplification bias was less than 3.5, which is small enough and appropriate for analysis of gene expressions in research about cell proliferation and tissue development. For other application in the medical field, the changes in the target gene expression may be much smaller than those in the cell development and tissue proliferation and thus this method may not be suitable. As the dynamic range in a DNA microarray analysis is only two to three orders, another quantitative analysis method with a larger dynamic range is required for medical applications.

A highly efficient technology for genome amplification from femtogram to microgram quantities is isothermal multiple displacement amplification (MDA) [74, 75]. MDA of the genome in a single cell has been developed. Its amplification efficiency is high enough for genome analysis, but amplification bias from single cell is still a problem as it could be as large as several hundred folds [74].

2.2 *Reusable Single-Cell cDNA Library on Beads Coupled with qPCR [16]*

For analyzing expression of multiple genes at a low level, it is very convenient if the produced cDNA library can be analyzed quantitatively and repeatedly. Since the most accurate quantitative method for the moment is qPCR, which does not require PCR amplification before the analysis, Taniguchi et al. have developed a method that uses a single-cell cDNA library on beads repeatedly coupled with qPCR. As the number of target molecules is very small, the loss of samples due to their adsorption on surfaces should be avoided. Accordingly, to avoid the sample loss, they used one tube to keep the single-cell cDNA library in throughout the experiment (see Fig. 2).

First, a cell is placed in a tube to be lysed, and the genomic DNA in the cell is digested with DNase, which is deactivated by heating. Magnetic beads with immobilized polyT probes are then added to the reaction mixture to capture mRNA with the probes. cDNA synthesis is carried out to produce a single-cell cDNA library on beads in the tube. The residual reagents inhibit qPCR and are removed completely from the tube while the cDNA library on beads is held in the tube. Primer pairs for gene A are then added to the tube for the first qPCR. After

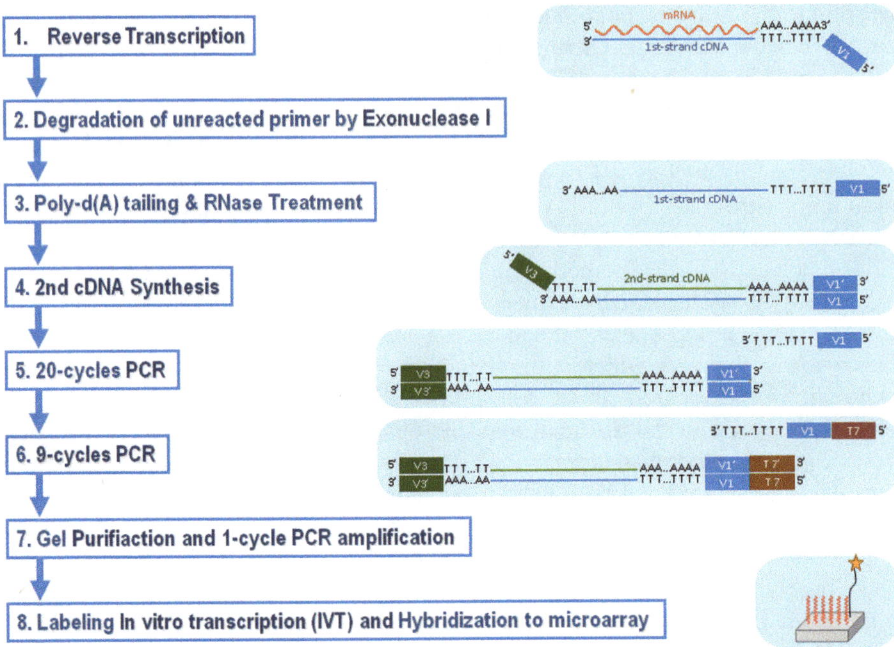

Fig. 2 Reusable single-cell cDNA library coupled with repeated qPCR method

that, the reaction products, as well as the reaction mixture, are completely removed from the tube. The second qPCR for gene B is then carried out and the same process is repeated. The reusability of a cDNA library on beads is limited by cDNA desorption from the bead surfaces due to thermal damage of the bead surfaces. To prevent this damage, Taniguchi et al. developed a low-temperature qPCR by adding hormamide into the reaction mix. The low-temperature PCR reduced the cDNA desorption rate from the bead surfaces to 2.8%/operation from 7.6%/ operation in the normal qPCR condition. It is possible to use a single-cell cDNA library on beads for qPCR more than ten times. It is also possible to analyze genes expressed at a level as low as ten copies in a single cell. Protein adsorption on bead surfaces usually occurs after several repeated uses of the cDNA library on beads, causing bead aggregation, which is not favorable for accurate qPCR. To prevent protein adsorption on beads and, therefore, prevent bead aggregation, 0.5% of MPC (2-methacryloyloxyethyl phosphorylcholine) polymer is added to the reaction mixture. MPC polymer is an excellent surfactant for dispersing the beads while not disturbing the PCR processes [76, 77].

Increasing mRNA-capturing efficiency and reverse-transcription efficiency is the key to precise gene expression analysis. Accordingly, various reverse transcriptases have been investigated, and these investigations concluded that Superscript III (Invitrogen) is the best from the viewpoint of efficiency as well as

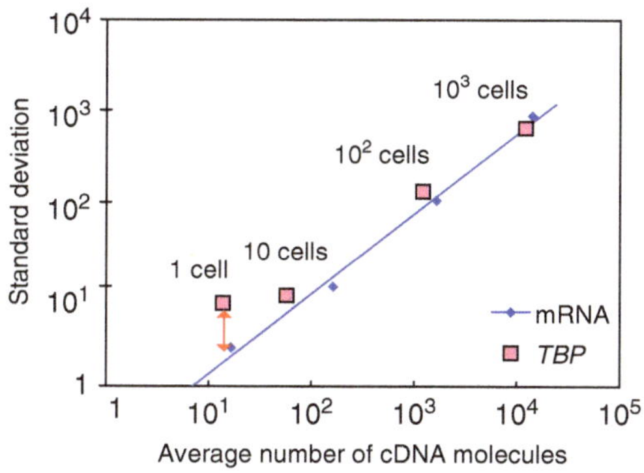

Fig. 3 Standard deviations of number of cDNA molecules for model samples and real samples

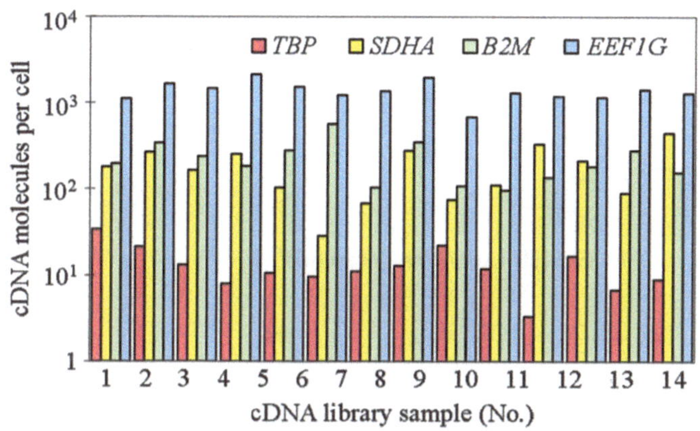

Fig. 4 Gene expressions for four housekeeping genes in single cells

less bead aggregation. The mRNA capturing efficiency depends on the density of the capturing probes in a reaction mix. One bead immobilizes 10^5 probes on the surface. The optimum bead number was around 10^7 for a 30-L reaction mix. The cDNA production efficiency was estimated with model mRNA to be over 85%.

The accuracy of the measurements was estimated with model samples. The noise levels were 5–15% of the observed cDNA amounts. The noise was mainly from random error induced in the handling procedures and in fluorescent measurements involved in qPCR. The standard deviations of number of cDNA molecules for various sample amounts are plotted in Fig. 3 for real cell samples as well as model mRNA samples. The standard deviations for the model samples linearly decrease with the number of cDNA molecules and approach Poisson noise. Those for cell samples fluctuated greatly, especially in the small-cell-number region, which may reflect the heterogeneity of cells. Gene expressions for four housekeeping genes in single cells are shown in Fig. 4. Fluctuations of gene expression levels among cells are observed. These results indicate that the cDNA-library-on-beads method is suitable for obtaining very accurate gene expression levels in single cells.

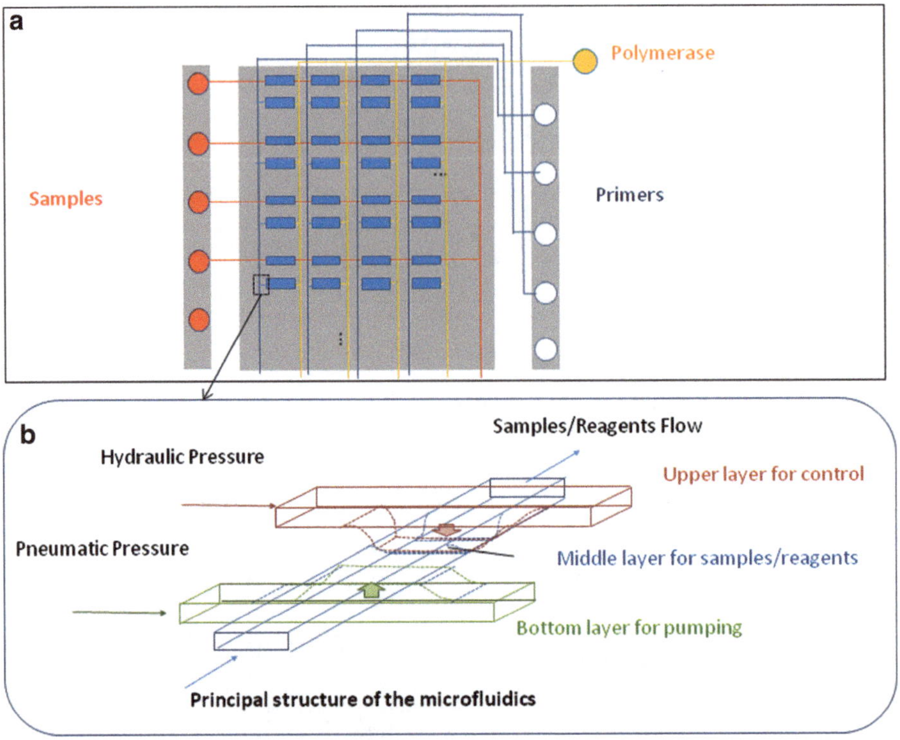

Fig. 5 Design and principle of IFC chip by Liu et al. [78]

2.3 Multiple Gene Expression Analysis with a Microfluidic Device [27, 51, 73, 78–82]

Quantitative PCR analysis of multiple gene expressions for single cells can be carried out with a microfluidic device coupled with cDNA amplification prior to the analysis when the amplification bias is not so large compared to the changes of gene expression levels in target phenomena. Although DNA array methods can analyze many gene expressions at the same time, it suffers from various drawbacks when applied to single cell setting. The quantitative accuracy is not so high because they use probe hybridization and the dynamic range for the microarray method is only two to three orders of magnitude. On the other hand, qPCR provides accurate and sensitive detection (detection of five to six copies of cDNA is possible) and a large dynamic range (six orders) data. However, one drawback of qPCR is that a sample has to be divided into many fractions for multiple gene expression analysis, and such division is detrimental to analyzing genes expressed at low levels. A device from Fluidigm Corporation performs qPCR on an integrated fluidic circuit chip (IFC chip) coupled with whole-cDNA amplification procedures [51]. A schematic view of the chip is shown in Fig. 5 [51, 78, 79, 83]. It consists of many micro-reaction chambers, sample as well as reagent-injection pathways, and hydraulic valves controlling liquid flows and pneumatic valves pumping liquids flows. The samples can be divided and mixed with many kinds of PCR primers in the micro-reaction chambers, and PCR reactions are carried out in these chambers simultaneously. The IFC chip therefore has large scalability in terms of number of samples and number of genes. Gene expressions of 96 genes for 96 single cells can be obtained in 4 h. As all the processes are automatically carried out, errors due to human handling are minimized. As the reaction volume is small (i.e., 7 nL), material and labor costs are greatly reduced. Although the number of gene species analyzable with an IFC chip is limited, this method realizes a simple and accurate multiple gene expression analysis for single cells.

Digital PCR based on a microfluidic device is another method which is suitable for accurate gene expression analysis [27]. In this method PCR bias can be avoided because quantification of cDNA is carried out by digital counting of micro-reaction chambers where single copy of cDNA is randomly injected to the individual chamber and PCR amplification is performed and detected by fluorescent detection. Digital PCR is accurate but the maximum number of molecules for quantization and number of samples are limited by the number of micro-reaction chambers. Droplet digital PCR is effective to increase the micro-reaction chamber without scaling up of size of the devices. This technology has been widely used recently, for example, to quantify DNA copy number [84] and to perform targeted PCR for high throughput sequencing [85]. However, application in single cell analysis remains to be explored.

Integration of all steps for single-cell processing in microfluidics is expected not only for research use but also has good potential for clinical use. Hansen et al. developed fully integrated microfluidic devices capable of cell capture, cell lysis, and RT-qPCR from 300 single cells per run [80]. This microfluidic format has an advantage in terms of correlation between a microscope image and gene expression and/or genome mutation [80]. Reduction of reaction volume makes it possible to reduce variation of genome amplification to avoid contaminant DNA templates and non-specific interaction between primers [86]. Marcy et al. developed a device with 60-nL reaction chambers for MDA from single bacterial genome and demonstrated single-cell amplification to 1.4×10^7 copies with geometric standard deviation of 19. A fully integrated microfluidic device also enables highly accurate and sensitive measurements, so that a single nucleotide variation in a single cell can be detected [81].

2.4 Digital Counting of cDNA Molecules with Next-Generation Sequencings

The progress of DNA sequencers is very fast and is creating new frontiers. Massive parallel DNA sequencer platforms have been used successfully to analyze gene expressions in many cells types [66–70, 87]. Recently, NGS has been applied to analyze gene expressions in single cells. As the amount of DNA required for sequencing is usually more than 200 ng, and the total amount of mRNA in a single cell is only 2 pg, global amplification of a cDNA library is necessary. Tang et al. analyzed gene expression in a single cell by cDNA sequencing coupled with global amplification for single cells [67]. Achieving global amplification without bias is a key requirement. Such amplification could be used to find new genes by sequencing at high coverage and depth; however, current quantitative accuracy in a single cell setting is not so good. This is because of the bias in global cDNA amplification as well as only part of the amplified products being sequenced. In many cases, relative change of gene expressions in cells in various states is the primary interest. To eliminate the amplification bias among cell samples in different states, Islam et al. introduced cell tags to discriminate cDNA originated from different cells, and their method is called "single-cell tagged reverse transcription" (STRT) [69]. Each single cell is placed in a chamber on a titer plate to be lysed, and cDNA synthesis is performed. cDNA fragments are prepared from the 5′ ends of mRNA for sequencing. Six base-tag-sequences are then introduced into cDNA at the 5′ termini of mRNA. After the cell tags are introduced into cDNA from single cells they are mixed to produce a pooled sample consisting of tagged cDNA. Global amplification is then carried out. Although there are biases among different cDNA species, there is no bias among the same cDNA species from various single cells because all the cDNA sequences are amplified at the same time in the reaction solution, except for the tag sequences. The amplified cDNA fragments are sequenced by NGS to determine the differences in gene expression levels of a number of reads in single cells. Since the cell tags are introduced in the reverse transcription processes, cell-to-cell errors coming from the amplification process are also mitigated.

More recently, Kivioja et al. reported a molecular-tag method, where a tag is attached to each cDNA molecule before amplification [71]. To cancel out PCR bias completely, the molecular tags for quantifying the initial number of mRNA molecules before amplification were sequenced. As a result, almost absolute quantization of mRNA in the sample was achieved by counting the random tags connected to mRNA at the 5′ end of the strands (Fig. 6).

3 Potential Applications of Single-Cell Analysis in the Clinic

Although single-cell analysis is still in its infant stage, it has great potential in clinical diagnosis. It provides precise information on the origins of diseases and the status of a whole system on the basis of classification of cells by cell activity and gene regulatory network. Much research on gene-regulatory-network analysis has been carried out with pooled samples [88, 89]. Although various biochemical components in a cell interact with each other within individual cells and in various environments, the cells also interact with each other in a tissue. It is thus necessary to understand inter-cellular networks as well as intra-cellular networks for a complete understanding of a disease state of the tissue.

A conceptual diagram of a network of genes and a network among cells is shown in Fig. 7. Cells in a tissue can be classified into several groups according to their gene expression patterns at single-cell resolution. They interact with each other

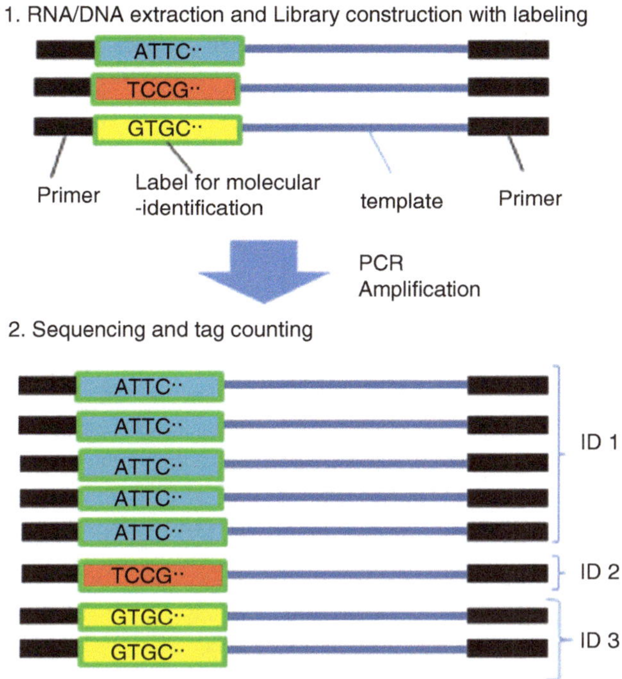

Fig. 6 Principle of absolute tag counting methods proposed by Kivioja et al.

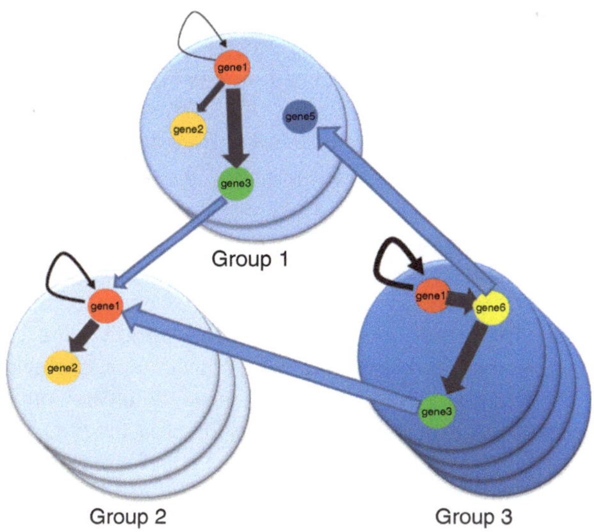

Fig. 7 Conceptual diagram of inter- and intra-cellular networks

through chemical molecules such as cytokines and chemokines, which affect the gene networks and, therefore, gene expression profiles in each group of cells. Gene expression analysis for single cells can determine active sub-networks of gene regulation in individual cells [90]. It also provides information on intercellular interactions [31, 33]. It is expected that such information is effective for determining the tissue state in an individual body. For example, this information may reveal the origins and properties of cancer cells (e.g., tissue type of cancer) and their susceptibility to drugs and the response of the immune system to the cancer cells and therapies.

The target cells of single-cell analysis include fetal cells [91–93], nucleated red blood cells (NRBC) [94], white blood cells [95, 96], circulating tumor cells [97–99], iPS cells [82], ES [100–102] cells, oocytes, and blastomeres [66, 67, 103]. Since the number of these cells for diagnoses is usually very small, NGS and microfluidic devices can be applied. The targets are genomic DNA or mRNA or micro RNA [101, 102] in single or several cells. For example, the numbers of CTC or NRBC in 1 mL of blood can be several copies of genome. It has been clarified that cells in a cancerous tissue exist in quite different states; this is called cellular heterogeneity [21, 104]. The cell population in various states gives important information on the characteristics of the cancerous tissue, which may be useful for development of new medicines. It has been pointed out that the gene expression patterns of pooled immune cells can be used to discover the origins of cancer and malignancy [95]. Furthermore, it has been clarified that macrophages differentiate into tumor-associated macrophages in order to adapt to the microenvironment and they are closely related to invasion, angiogenesis, immune-suppression, and metastasis [96]. For effective immune therapies, it is useful to analyze single cells to find tumor-associated immunocytes and to characterize them by gene expressions.

The analysis of transcriptomes in single cells is more effective for predicting prognosis than the analysis to determine pathological grade. Dalerba et al. analyzed transcriptomes in colon-tumor single cells [20]. They analyzed primary human normal colon and colon cancer epithelia. They demonstrated that the transcriptional diversity of cancer tissues is explained not only by clonal genetic heterogeneity but also by multi-lineage differentiation.

It is becoming important to analyze not only gene expressions but also genomic DNA of single cells, although such analyses are rather difficult for the moment. Many genomic DNA sequencing analyses on pooled genomic samples have been carried out (mainly for exome parts where various proteins are coded). As only a small fraction of exomes could be extracted from samples, a large number of cells are required for the analysis. Although it is not possible to carry out exome analysis on single cells at present, it may be possible in the next few years. Several groups have tried genomic DNA sequencing from single cells. Single-cell genome analysis and methylation analysis are very promising for determining lineages and malignancy of tumor cells and determining effective drugs in accordance with positions of mutations. In the past, the analysis of copy numbers for special genes is usually carried out by FISH. Now NGS can be used for this analysis. Recently, Navin et al. [21] reported a genome-copy-number analysis on cancerous single cells by using a next-generation DNA sequencer. They separated nuclei by a cell sorter and amplified the whole genomic DNA to produce template for NGS. They found copy-number profiles for every 54 kB by using sequencing data covered 6% of the whole genome in cancerous cells. The cells could be classified into four sub classes that coincide with those of flow-cytometry analysis. Moreover, they observed quasi-doublet cell groups that could be observed only by single-cell sequencing analysis. Further analysis of cancerous cells indicated that the cancerous tissue originated from cancer stem cells.

Conventional experimental methods for sequencing the human genome are limited by inevitable mixing of alleles or haplotypes. Fan et al. developed a microfluidic device capable of separating and amplifying homologous copies of each chromosome from a single human cell [81] by using MDA. Knowledge of complete haplotypes of individuals will be important for personal medicine for treating autoimmune disease and so on and for prognosis of transplantation by determining human-leukocyte-antigen (HLA) haplotypes. If a totally integrated device capable of determining genomes and transcriptomes in a single cell were developed, it might be able to clarify direct relations between genotype and phenotype.

Single-cell analysis for inter-cellular interaction has been used to improve effectiveness of T-cell therapy by extraction of subpopulations among tumor antigen-specific cytotoxic T lymphocytes (CTLs) on the basis of fluorescent imaging of secreted cytokines from single cells [31]. Ma et al. developed microfluidics capable of quantifying more than 10 proteins (cytokines) in 1,040 chambers with single cells. Moreover, Varadarajan et al. analyzed correlations between functionality of cells and cytokine secretion at single-cell resolution. They observed poor correlation between cytolysis and IFN-γ, although secretion of IFN-γ is widely used as a marker of cytolysis [33].

To sum up, it is concluded that, although it is still in its infancy and few tools have been developed for it, single-cell analysis will contribute to the new era of medical application of analysis in the future.

Acknowledgments This research was supported by the Strategic International Cooperative Program of the Japan Science and Technology Agency. M. Shirai, T. Taniguchi, and H. Kambara are employees in Hitachi Central Research Laboratory.

References

1. Margulies M, Egholm M, Altman WE, Attiya S, Bader JS, Bemben LA et al (2005) Genome sequencing in microfabricated high-density picolitre reactors. Nature 437:376–380
2. Mardis ER (2008) Next-generation DNA sequencing methods. Annu Rev Genomics Hum Genet 9:387–402
3. Ozsolak F, Platt AR, Jones DR, Reifenberger JG, Sass LE, McInerney P et al (2009) Direct RNA sequencing. Nature 461(7265):814–818

4. Rothberg JM, Hinz W, Rearick TM, Schultz J, Mileski W, Davey M et al (2011) An integrated semiconductor device enabling non-optical genome sequencing. Nature 475(7356):348–352
5. Shendure J, Ji H (2008) Next-generation DNA sequencing. Nat Biotechnol 26(10):1135–1145
6. Clark MJ, Chen R, Lam HY, Karczewski KJ, Euskirchen G, Butte AJ et al (2011) Performance comparison of exome DNA sequencing technologies. Nat Biotechnol 29(10):908–914
7. Lam HY, Clark MJ, Chen R, Natsoulis G, O'Huallachain M, Dewey FE et al (2011) Performance comparison of whole-genome sequencing platforms. Nat Biotechnol 30(1):78–82
8. Li J, Gao F, Li N, Li S, Yin G, Tian G et al (2009) An improved method for genome wide DNA methylation profiling correlated to transcription and genomic instability in two breast cancer cell lines. BMC Genomics 10:223
9. Liu S, Zhou Z, Lu J, Sun F, Wang S, Liu H et al (2011) Generation of genome-scale gene-associated SNPs in catfish for the construction of a high-density SNP array. BMC Genomics 12:53
10. Ruan X, Ruan Y (2012) Genome wide full-length transcript analysis using 5′ and 3′ paired-end-tag next generation sequencing (RNA-PET). Methods Mol Biol 809:535–562
11. Wheeler DA, Srinivasan M, Egholm M, Shen Y, Chen L, McGuire A et al (2008) The complete genome of an individual by massively parallel DNA sequencing. Nature 452 (7189):872–876
12. Maeda N, Nishiyori H, Nakamura M, Kawazu C, Murata M, Sano H et al (2008) Development of a DNA barcode tagging method for monitoring dynamic changes in gene expression by using an ultra high-throughput sequencer. Biotechniques 45(1):95–97
13. Elowitz MB (2002) Stochastic gene expression in a single cell. Science 297(5584): 1183–1186
14. Gautreau L, Boudil A, Pasqualetto V, Skhiri L, Grandin L, Monteiro M et al (2010) Gene coexpression analysis in single cells indicates lymphomyeloid copriming in short-term hematopoietic stem cells and multipotent progenitors. J Immunol 184(9):4907–4917
15. Hartmann CH, Klein CA (2006) Gene expression profiling of single cells on large-scale oligonucleotide arrays. Nucleic Acids Res 34(21):e143
16. Taniguchi K, Kajiyama T, Kambara H (2009) Quantitative analysis of gene expression in a single cell by qPCR. Nat Methods 6(7):503–506
17. Wang D, Bodovitz S (2010) Single cell analysis: the new frontier in 'omics'. Trends Biotechnol 28(6):281–290
18. Singh DK, Ku C-J, Wichaidit C, Steininger RJ, Wu LF, Altschuler SJ (2010) Patterns of basal signaling heterogeneity can distinguish cellular populations with different drug sensitivities. Mol Syst Biol 6:369–378
19. Batchelor E, Loewer A, Mock C, Lahav G (2011) Stimulus-dependent dynamics of p53 in single cells. Mol Syst Biol 7:488–15
20. Dalerba P, Kalisky T, Sahoo D, Rajendran PS, Rothenberg ME, Leyrat AA et al (2011) Single-cell dissection of transcriptional heterogeneity in human colon tumors. Nat Biotechnol 29(12):1120–1127
21. Navin N, Kendall J, Troge J, Andrews P, Rodgers L, McIndoo J et al (2011) Tumour evolution inferred by single-cell sequencing. Nature 472(7341):90–94
22. April CS, Fan JB (2011) Gene expression profiling in formalin-fixed, paraffin-embedded tissues using the whole-genome DASL assay. Methods Mol Biol 784:77–98
23. Beisel C, Paro R (2009) Dissection of gene regulatory networks in embryonic stem cells by means of high-throughput sequencing. Biol Chem 390(11):1139–1144
24. Bibikova M, Laurent LC, Ren B, Loring JF, Fan JB (2008) Unraveling epigenetic regulation in embryonic stem cells. Cell Stem Cell 2(2):123–134
25. Rieger C, Poppino R, Sheridan R, Moley K, Mitra R, Gottlieb D (2007) Polony analysis of gene expression in ES cells and blastocysts. Nucleic Acids Res 35(22):e151

26. Bendall SC, Simonds EF, Qiu P, Amir EaD, Krutzik PO, Finck R et al (2011) Single-cell mass cytometry of differential immune and drug responses across a human hematopoietic continuum. Science 332(6030):687–696
27. Warren L, Bryder D, Weissman IL, Quake SR (2006) Transcription factor profiling in individual hematopoietic progenitors by digital RT-PCR. Proc Natl Acad Sci USA 103(47):17807–17812
28. Femino AM (1998) Visualization of single RNA transcripts in situ. Science 280 (5363):585–590
29. Flatz L, Roychoudhuri R, Honda M, Filali-Mouhim A, Goulet JP, Kettaf N et al (2011) Single-cell gene-expression profiling reveals qualitatively distinct CD8 T cells elicited by different gene-based vaccines. Proc Natl Acad Sci USA 108(14):5724–5729
30. Itzkovitz S, van Oudenaarden A (2011) Validating transcripts with probes and imaging technology. Nat Methods 8(4s):S12–S19
31. Ma C, Fan R, Ahmad H, Shi Q, Comin-Anduix B, Chodon T et al (2011) A clinical microchip for evaluation of single immune cells reveals high functional heterogeneity in phenotypically similar T cells. Nat Med 17(6):738–743
32. Orth JD, Kohler RH, Foijer F, Sorger PK, Weissleder R, Mitchison TJ (2011) Analysis of mitosis and antimitotic drug responses in tumors by in vivo microscopy and single-cell pharmacodynamics. Cancer Res 71(13):4608–4616
33. Varadarajan N, Julg B, Yamanaka YJ, Chen H, Ogunniyi AO, McAndrew E et al (2011) A high-throughput single-cell analysis of human CD8+ T cell functions reveals discordance for cytokine secretion and cytolysis. J Clin Invest 121(11):4322–4331
34. Weibrecht I, Grundberg I, Nilsson M, Soderberg O (2011) Simultaneous visualization of both signaling cascade activity and end-point gene expression in single cells. PLoS One 6(5):e20148
35. Rubakhin SS, Romanova EV, Nemes P, Sweedler JV (2011) Profiling metabolites and peptides in single cells. Nat Methods 8(Suppl 4):S20–S29
36. Masujima T (2009) Live single-cell mass spectrometry. Anal Sci 25(8):953–960
37. Mizuno H, Tsuyama N, Harada T, Masujima T (2008) Live single-cell video-mass spectrometry for cellular and subcellular molecular detection and cell classification. J Mass Spectrom 43(12):1692–1700
38. Tsuyama N, Mizuno H, Masujima T (2011) Mass spectrometry for cellular and tissue analyses in a very small region. Anal Sci 27(2):163–170
39. Agrawal A, Zhang C, Byassee T, Tripp RA, Nie S (2006) Counting single native biomolecules and intact viruses with color-coded nanoparticles. Anal Chem 78(4):1061–1070
40. Duncan RR (2006) Fluorescence lifetime imaging microscopy (FLIM) to quantify protein-protein interactions inside cells. Biochem Soc Trans 34(Pt 5):679–682
41. Huang B, Wu H, Bhaya D, Grossman A, Granier S, Kobilka BK et al (2007) Counting low-copy number proteins in a single cell. Science 315(5808):81–84
42. Raj A, Peskin CS, Tranchina D, Vargas DY, Tyagi S (2006) Stochastic mRNA synthesis in mammalian cells. PLoS Biol 4(10):e309
43. Taniguchi Y, Choi PJ, Li GW, Chen H, Babu M, Hearn J et al (2010) Quantifying E. coli proteome and transcriptome with single-molecule sensitivity in single cells. Science 329 (5991):533–538
44. Kurimoto K, Saitou M (2010) Single-cell cDNA microarray profiling of complex biological processes of differentiation. Curr Opin Genet Dev 20(5):470–477
45. Kurimoto K (2006) An improved single-cell cDNA amplification method for efficient high-density oligonucleotide microarray analysis. Nucleic Acids Res 34(5):e42
46. Seshi B, Kumar S, King D (2003) Multilineage gene expression in human bone marrow stromal cells as evidenced by single-cell microarray analysis. Blood Cells Mol Dis 31 (2):268–285

47. Kamme F, Salunga R, Yu J, Tran DT, Zhu J, Luo L et al (2003) Single-cell microarray analysis in hippocampus CA1: demonstration and validation of cellular heterogeneity. J Neurosci 23(9):3607–3615
48. Chiang MK, Melton DA (2003) Single-cell transcript analysis of pancreas development. Dev Cell 4(3):383–393
49. Brail LH, Jang A, Billia F, Iscove NN, Klamut HJ, Hill RP (1999) Gene expression in individual cells: analysis using global single cell reverse transcription polymerase chain reaction (GSC RT-PCR). Mutat Res 406(2–4):45–54
50. Citri A, Pang ZP, Sudhof TC, Wernig M, Malenka RC (2012) Comprehensive qPCR profiling of gene expression in single neuronal cells. Nat Protoc 7(1):118–127
51. Spurgeon SL, Jones RC, Ramakrishnan R (2008) High throughput gene expression measurement with real time PCR in a microfluidic dynamic array. PLoS One 3(2):e1662
52. Huang Q, Lin B, Liu H, Ma X, Mo F, Yu W et al (2011) RNA-Seq analyses generate comprehensive transcriptomic landscape and reveal complex transcript patterns in hepatocellular carcinoma. PLoS One 6(10):e26168
53. Chang ST, Sova P, Peng X, Weiss J, Law GL, Palermo RE et al (2011) Next-generation sequencing reveals HIV-1-mediated suppression of T cell activation and RNA processing and regulation of noncoding RNA expression in a CD4+ T cell line. MBio 2(5):e00134–11
54. Asmann YW, Wallace MB, Thompson EA (2008) Transcriptome profiling using next-generation sequencing. Gastroenterology 135(5):1466–1468
55. Baxter I, Hashimoto S-i, Qu W, Ahsan B, Ogoshi K, Sasaki A et al (2009) High-resolution analysis of the 5′-end transcriptome using a next generation DNA sequencer. PLoS One 4(1):e4108
56. Cloonan N, Forrest AR, Kolle G, Gardiner BB, Faulkner GJ, Brown MK et al (2008) Stem cell transcriptome profiling via massive-scale mRNA sequencing. Nat Methods 5(7):613–619
57. Collins LJ, Biggs PJ, Voelckel C, Joly S (2008) An approach to transcriptome analysis of non-model organisms using short-read sequences. Genome Inform 21:3–14
58. Elling AA, Deng XW (2009) Next-generation sequencing reveals complex relationships between the epigenome and transcriptome in maize. Plant Signal Behav 4(8):760–762
59. Fullwood MJ, Wei CL, Liu ET, Ruan Y (2009) Next-generation DNA sequencing of paired-end tags (PET) for transcriptome and genome analyses. Genome Res 19(4):521–532
60. Hanriot L, Keime C, Gay N, Faure C, Dossat C, Wincker P et al (2008) A combination of LongSAGE with Solexa sequencing is well suited to explore the depth and the complexity of transcriptome. BMC Genomics 9:418
61. Kuchenbauer F, Morin RD, Argiropoulos B, Petriv OI, Griffith M, Heuser M et al (2008) In-depth characterization of the microRNA transcriptome in a leukemia progression model. Genome Res 18(11):1787–1797
62. Maningat PD, Sen P, Rijnkels M, Sunehag AL, Hadsell DL, Bray M et al (2009) Gene expression in the human mammary epithelium during lactation: the milk fat globule transcriptome. Physiol Genomics 37(1):12–22
63. Morrissy AS, Morin RD, Delaney A, Zeng T, McDonald H, Jones S et al (2009) Next-generation tag sequencing for cancer gene expression profiling. Genome Res 19 (10):1825–1835
64. Porter S, Olson NE, Smith T (2009) Analyzing gene expression data from microarray and next-generation DNA sequencing transcriptome profiling assays using GeneSifter analysis edition. Curr Protoc Bioinformatics Chapter 7:Unit 7.14.7 1–35
65. Hashimoto S, Qu W, Ahsan B, Ogoshi K, Sasaki A, Nakatani Y et al (2009) High-resolution analysis of the 5'-end transcriptome using a next generation DNA sequencer. PLoS One 4(1): e4108
66. Tang F, Barbacioru C, Nordman E, Li B, Xu N, Bashkirov VI et al (2010) RNA-Seq analysis to capture the transcriptome landscape of a single cell. Nat Protoc 5(3):516–535
67. Tang F, Barbacioru C, Wang Y, Nordman E, Lee C, Xu N et al (2009) mRNA-Seq whole-transcriptome analysis of a single cell. Nat Methods 6(5):377–382

68. Lao KQ, Tang F, Barbacioru C, Wang Y, Nordman E, Lee C et al (2009) mRNA-sequencing whole transcriptome analysis of a single cell on the SOLiD system. J Biomol Tech 20(5):266–271
69. Islam S, Kjallquist U, Moliner A, Zajac P, Fan JB, Lonnerberg P et al (2011) Characterization of the single-cell transcriptional landscape by highly multiplex RNA-seq. Genome Res 21(7):1160–1167
70. Linnarsson S (2010) Recent advances in DNA sequencing methods – general principles of sample preparation. Exp Cell Res 316(8):1339–1343
71. Kivioja T, Vaharautio A, Karlsson K, Bonke M, Enge M, Linnarsson S et al (2011) Counting absolute numbers of molecules using unique molecular identifiers. Nat Methods 9(1):72–74
72. Brady G, Iscove NN (1993) Construction of cDNA libraries from single cells. Methods Enzymol 225:611–623
73. Unger MA, Chou HP, Thorsen T, Scherer A, Quake SR (2000) Monolithic microfabricated valves and pumps by multilayer soft lithography. Science 288(5463):113–116
74. Lasken RS (2007) Single-cell genomic sequencing using multiple displacement amplification. Curr Opin Microbiol 10(5):510–516
75. Zhang K, Martiny AC, Reppas NB, Barry KW, Malek J, Chisholm SW et al (2006) Sequencing genomes from single cells by polymerase cloning. Nat Biotechnol 24(6):680–686
76. Ishihara K, Takai M (2009) Bioinspired interface for nanobiodevices based on phospholipid polymer chemistry. J R Soc Interface 6(Suppl 3):S279–S291
77. Xu Y, Takai M, Ishihara K (2009) Protein adsorption and cell adhesion on cationic, neutral, and anionic 2-methacryloyloxyethyl phosphorylcholine copolymer surfaces. Biomaterials 30(28):4930–4938
78. Liu J, Hansen C, Quake SR (2003) Solving the "world-to-chip" interface problem with a microfluidic matrix. Anal Chem 75(18):4718–4723
79. Thorsen T, Maerkl SJ, Quake SR (2002) Microfluidic large-scale integration. Science 298(5593):580–584
80. White AK, VanInsberghe M, Petriv OI, Hamidi M, Sikorski D, Marra MA et al (2011) High-throughput microfluidic single-cell RT-qPCR. Proc Natl Acad Sci USA 108(34):13999–14004
81. Fan HC, Wang J, Potanina A, Quake SR (2011) Whole-genome molecular haplotyping of single cells. Nat Biotechnol 29(1):51–57
82. Narsinh KH, Sun N, Sanchez-Freire V, Lee AS, Almeida P, Hu S et al (2011) Single cell transcriptional profiling reveals heterogeneity of human induced pluripotent stem cells. J Clin Invest 121(3):1217–1221
83. Unger MA (2000) Monolithic microfabricated valves and pumps by multilayer soft lithography. Science 288(5463):113–116
84. Hindson BJ, Ness KD, Masquelier DA, Belgrader P, Heredia NJ, Makarewicz AJ et al (2011) High-throughput droplet digital PCR system for absolute quantitation of DNA copy number. Anal Chem 83(22):8604–8610
85. Tewhey R, Warner JB, Nakano M, Libby B, Medkova M, David PH et al (2009) Microdroplet-based PCR enrichment for large-scale targeted sequencing. Nat Biotechnol 27(11):1025–1031
86. Marcy Y, Ishoey T, Lasken RS, Stockwell TB, Walenz BP, Halpern AL et al (2007) Nanoliter reactors improve multiple displacement amplification of genomes from single cells. PLoS Genet 3(9):1702–1708
87. Eberwine J, Bartfai T (2011) Single cell transcriptomics of hypothalamic warm sensitive neurons that control core body temperature and fever response signaling asymmetry and an extension of chemical neuroanatomy. Pharmacol Ther 129(3):241–259
88. Suzuki H, Forrest AR, van Nimwegen E, Daub CO, Balwierz PJ, Irvine KM et al (2009) The transcriptional network that controls growth arrest and differentiation in a human myeloid leukemia cell line. Nat Genet 41(5):553–562
89. Akalin A, Fredman D, Arner E, Dong X, Bryne J, Suzuki H et al (2009) Transcriptional features of genomic regulatory blocks. Genome Biol 10(4):R38

90. Trott J, Hayashi K, Surani A, Babu MM, Martinez-Arias A (2012) Dissecting ensemble networks in ES cell populations reveals micro-heterogeneity underlying pluripotency. Mol Biosyst 8:744–752
91. Hahn S, Jackson LG, Kolla V, Mahyuddin AP, Choolani M (2009) Noninvasive prenatal diagnosis of fetal aneuploidies and Mendelian disorders: new innovative strategies. Expert Rev Mol Diagn 9(6):613–621
92. Lo YM, Chiu RW (2008) Noninvasive prenatal diagnosis of fetal chromosomal aneuploidies by maternal plasma nucleic acid analysis. Clin Chem 54(3):461–466
93. Lun FM, Chiu RW, Allen Chan KC, Yeung Leung T, Kin Lau T, Dennis Lo YM (2008) Microfluidics digital PCR reveals a higher than expected fraction of fetal DNA in maternal plasma. Clin Chem 54(10):1664–1672
94. Lo YM, Tsui NB, Chiu RW, Lau TK, Leung TN, Heung MM et al (2007) Plasma placental RNA allelic ratio permits noninvasive prenatal chromosomal aneuploidy detection. Nat Med 13(2):218–223
95. Honda M, Sakai Y, Yamashita T, Sakai A, Mizukoshi E, Nakamoto Y et al (2010) Differential gene expression profiling in blood from patients with digestive system cancers. Biochem Biophys Res Commun 400(1):7–15
96. Lewis CE, Pollard JW (2006) Distinct role of macrophages in different tumor microenvironments. Cancer Res 66(2):605–612
97. Solmi R, De Sanctis P, Zucchini C, Ugolini G, Rosati G, Del Governatore M et al (2004) Search for epithelial-specific mRNAs in peripheral blood of patients with colon cancer by RT-PCR. Int J Oncol 25(4):1049–1056
98. Li G, Passebosc-Faure K, Gentil-Perret A, Lambert C, Genin C, Tostain J (2005) Cadherin-6 gene expression in conventional renal cell carcinoma: a useful marker to detect circulating tumor cells. Anticancer Res 25(1A):377–381
99. Smith B, Selby P, Southgate J, Pittman K, Bradley C, Blair GE (1991) Detection of melanoma cells in peripheral blood by means of reverse transcriptase and polymerase chain reaction. Lancet 338(8777):1227–1229
100. Tang F, Barbacioru C, Bao S, Lee C, Nordman E, Wang X et al (2010) Tracing the derivation of embryonic stem cells from the inner cell mass by single-cell RNA-seq analysis. Cell Stem Cell 6(5):468–478
101. Tang F, Hajkova P, O'Carroll D, Lee C, Tarakhovsky A, Lao K et al (2008) MicroRNAs are tightly associated with RNA-induced gene silencing complexes in vivo. Biochem Biophys Res Commun 372(1):24–29
102. Tang F (2006) MicroRNA expression profiling of single whole embryonic stem cells. Nucleic Acids Res 34(2):e9
103. Tang F, Barbacioru C, Nordman E, Bao S, Lee C, Wang X et al (2011) Deterministic and stochastic allele specific gene expression in single mouse blastomeres. PLoS One 6(6):e21208
104. Navin N, Hicks J (2011) Future medical applications of single-cell sequencing in cancer. Genome Med 3(5):31

Top Curr Chem (2014) 336: 117–138
DOI: 10.1007/128_2012_353

Published online: 12 August 2012

Mass Spectrometry in High-Throughput Clinical Biomarker Assays: Multiple Reaction Monitoring

Carol E. Parker, Dominik Domanski, Andrew J. Percy, Andrew G. Chambers, Alexander G. Camenzind, Derek S. Smith, and Christoph H. Borchers

Abstract Clinical biomarker discovery, verification, and validation are facilitated by the latest technological advances in mass spectrometry. It is now possible to analyze simultaneously group of tens or hundreds of biomarkers in a blood sample using multiple reaction monitoring (MRM), a tandem mass spectrometric method. However, these newly-developed methods face new challenges, including standardization, calibration, and the determination of analytical and biological variation. Here we illustrate the background, pre-analytical sample preparation, and biomarker assay development using an MRM-mass spectrometric method. In addition, special attention is given to future standardization methods to enable widespread use of the technology.

Keywords Clinical proteomics · Multiple reaction monitoring (MRM) · Multiplex · Proteomics · Quantitation · Selected reaction monitoring (SRM)

Contents

1 Introduction 119
 1.1 Instrumentation 119
 1.2 MRM-Based Quantitation 119
 1.3 Development of an MRM Method 122

C.E. Parker, D. Domanski, A.J. Percy, A.G. Chambers, A.G. Camenzind, and D.S. Smith
University of Victoria, University of Victoria – Genome BC Proteomics Centre, #3101-4464 Markham Street, Vancouver Island Technology Park, Victoria, BC V8Z 7X8, Canada

C.H. Borchers (✉)
University of Victoria, University of Victoria – Genome BC Proteomics Centre, #3101-4464 Markham Street, Vancouver Island Technology Park, Victoria, BC V8Z 7X8, Canada

University of Victoria, Department of Biochemistry and Microbiology, Petch Building, Room 207, 3800 Finnerty Rd., Victoria, BC V8P 5C2, Canada
e-mail: christoph@proteincentre.com

2 Current Research in MRM-Based Clinical Proteomics 126
2.1 Targeted MRM-Based Quantitation 126
2.2 Multiplexed MRM for Biomarker Discovery 128
2.3 MRM for Biomarker Verification 128
2.4 MRM for Pathway Analysis 129
2.5 Stable Isotope Standards and Capture by Anti-peptide Antibodies 130
2.6 Matrix-Assisted Laser-Desorption/Ionization-Multiple Reaction Monitoring 130
3 Statistics and Data Analysis 131
4 Need for Harmonization 133
5 Need for Standardization 134
References 135

List of Abbreviations

APO4	Apolipoprotein A4
AUC	Area under the curve
CE	Collision energy
CID	Collision-induced dissociation
CV	Coefficient of variation
CVD	Cardiovascular disease
DDA	Data-dependent acquisition
EDRN	Early Detection Research Network
EDTA	Ethylenediamine tetraacetic acid
ESI	Electrospray ionization
HUPO	Human Proteome Organization
iMALDI	Immuno-matrix-assisted laser-desorption/ionization
iTRAQ™	Isobaric tag for relative and absolute quantitation
LC	Liquid chromatography
LOD	Limit of detection
LOQ	Limit of quantitation
MALDI	Matrix-assisted laser-desorption/ionization
MMP8	Matrix metalloproteinase-8
MRM	Multiple reaction monitoring
MS/MS	Tandem mass spectrometry
PSA	Prostate specific antigen
Q1 (Q2, Q3)	Quadrupole 1 (quadrupole 2, quadrupole 3)
QUAD	The Quantitative Assay Database
ROC	Receiver operator curve (or receiver operator characteristic)
SIS peptides	Stable-isotope-labeled internal standard peptides
SISCAPA	Stable isotope standards and capture by anti-peptide antibodies
SRM	Selected reaction monitoring
UHPLC	Ultrahigh-performance liquid chromatography

1 Introduction

1.1 Instrumentation

Multiple reaction monitoring (MRM), also called selected reaction monitoring (SRM), is a mass spectrometric scanning technique that was developed in the 1970s on magnetic sector instruments and on triple quadrupole instruments [1]. MRM can be thought of as a combination of selected ion monitoring and tandem mass spectrometry (MS/MS), a technique named because it involves two mass selection steps. In a triple-quadrupole instrument, the first and third quadrupoles (Q1 and Q3) perform the mass selection, acting as mass filters; the second (rf-only) quadrupole (Q2), where the fragmentation occurs, acts as collision cell. The reaction of the precursor to the product ion is measured, giving the technique its name. Although in early versions of this technique the mass spectrometers were not scanned [2], the added information generated by having a peak profile (instead of simply having an ion current value) was soon recognized [3] and narrow-range scanning was added to both selection ion monitoring and SRM experiments. Thus, the current version of MRM can be thought of as a type of product ion scan [2], with the precursor ion being selected and Q3 being scanned over a very narrow mass range.

The schematic of an MRM experiment is given in Fig. 1. After ionization in the source, the first stage of mass selection is carried out in Q1 and involves selection of a precursor mass. Next, fragmentation of the selected precursor ion is carried out in the second quadrupole (Q2). This is usually carried out by collision-induced dissociation (CID), where fragmentation of the precursor ion is caused by collision with a neutral gas (air, nitrogen, argon, etc.). The degree of fragmentation is usually controlled by changing the energetics of the collision (collision energy, CE) rather than changing the collision gas pressure. In practice, the collision gas pressure is kept constant, and the velocity of the precursor ion is changed. To change the velocity, fairly low voltages (10–35 eV) are used. Larger ions take more energy to fragment than smaller ions, and most software systems use an equation to estimate the collision energy required, as a function of the molecular weight of the precursor ion. This works fairly well for data-dependent acquisition (DDA) (where one really has no choice but to use this approximation), but for MRM where the target ions are known and standards are available, we have achieved up to an 11.4-fold enhancement by empirical tuning of the collision energy [4, 5].

1.2 MRM-Based Quantitation

The MRM technique provides four degrees of specificity if stable-isotope-labeled internal standard peptides (SIS peptides) are used. These are 1) the precursor mass, 2) the product ion mass, and 3) the ratios of the relative abundances of the product ions for the endogenous peptide, all of which must match those of the SIS standard.

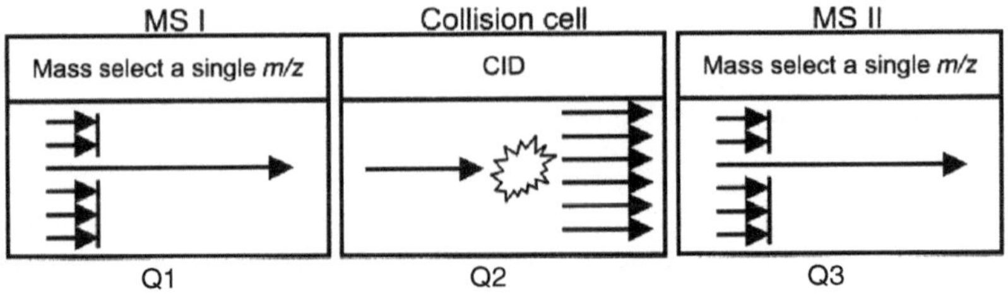

Fig. 1 Schematic of an MRM experiment on a triple-quadrupole mass spectrometer. Adapted from [58], with permission

Also, since MRM is normally carried out with the instrument coupled to a liquid chromatography (LC), all of these peaks must co-elute (Fig. 2), thus providing a fourth degree of specificity.

When used in conjunction with stable isotope-labeled (^{13}C or ^{15}N) internal standards, both the precursor and fragment ion peak from the sample and the precursor and fragment ion peaks from the standard must coelute (because the SIS standard is chemically identical to the endogenous material it also has the same precursor/fragment ion ratios). In addition to enhanced specificity, this two-stage ion separation also reduces background chemical noise and can lead to enhanced detection sensitivity for the targeted analyte, compared to full-scan DDA modes. We should point out that deuterium-labeled peptides are not considered to be SIS peptides because while ^{13}C-labeled or ^{15}N-labeled standard peptides are chemically identical to their endogenous counterparts [6] and therefore have identical fragmentation patterns (taking into account the mass shifts) and identical retention times, deuterated peptides have different properties than their hydrogen-labeled counterparts [7, 8], including different reversed-phase LC retention times, so using them would negate some of the advantages of using stable-labeled internal standards.

The combination of specificity and sensitivity of MRM with SIS peptides has made this technique the "gold standard" for quantitation of small molecules [9], including peptides. Because the mass spectrometric resolution and sensitivity is higher for peptides than that of intact proteins, most proteomics work is still done by "bottom-up" proteomics. Concentration information can be inferred by measuring the concentration of peptides unique to and characteristic of the parent protein ("proteotypic" peptides). Another important consideration for MS/MS-based methods such as MRM is that, because the collision energy is distributed throughout the peptide, there is a size limitation to peptides that can be fragmented by low-energy CID (usually <2,500 Da). This is another reason why "bottom up" methods are used for most peptide/protein quantitation [10].

There are several "caveats" which must be mentioned here. The first is the width of the precursor mass window. While you might be able to accurately set the center of the precursor mass window, there is an instrument-specific width to this window, which can be several Da wide. Any ions within this window will pass through Q1 and reach the collision cell. The presence of these ions, sometimes called

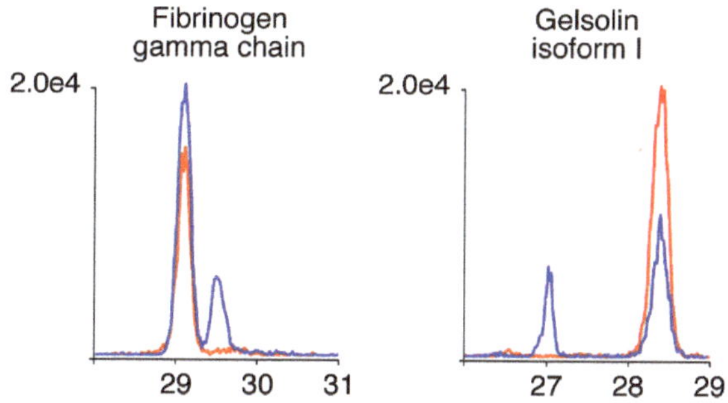

Fig. 2 The alignment of the SIS (*red*) with the endogenous peptide (*blue*) provides clear identification of the endogenous peptide. Adapted from [5], with permission

"chimeric" ions, can lead to mixed MS/MS spectra. This can result in uninterpretable spectra in DDA analysis and possible interferences in MRM analysis. In MRM this would be observable as an incorrect precursor/product ion ratio for the standard and the target analyte. If this is noted, a replacement peptide would usually be selected for the final MRM method.

1.2.1 Pre-analytical Considerations

Other potential errors in the accuracy of the MRM results come from upstream sample collection and treatment procedures, the "pre-analytical" variability. This is an important consideration that clinical chemists face more frequently than standard "analytical chemists," which is why clinical chemists have separated out this part of the process and given it this special name. Biological fluids, particularly plasma, are dynamic matrices. Special care must be taken in sample collection and preservation to deactivate enzymes (usually through the addition of protease inhibitors) and not to release proteolytic enzymes that could alter the composition of the sample. Repeated freeze–thaw cycles should be avoided. Although it is still used in proteomics, the process of generating serum from plasma involves coagulation which is inherently variable. This has precluded the development of serum collection tubes. Several studies have been performed comparing different types of plasma collection tubes to minimize this "pre-analytical" variability [11, 12], and the conclusion of our recent study is that following the protocols recommended by HUPO and EDRN [13, 14] and using K2-EDTA tubes gives good results for protein biomarker studies, but that special care must be taken to avoid platelet activation, particularly if cytokines are to be measured [12]. Collection of a separate aliquot in CTAD tubes was recommended for cytokine quantitation – thus the type of sample tube recommended depends on the particular application.

Since MRM-based quantitation is usually carried out on peptides, the digestion step can be another major source of variability. Urea or deoxycholate are normally used for the digestion, although we have demonstrated that the completeness of the digestion process (and the optimal digestion conditions) vary with the target protein and even with the target peptide [15]. Our study has indicated that for most proteins, or for a mixture of proteins (for example, for shotgun proteomics), deoxycholate digestion is preferable. There are actually several factors at work here – one is whether the digestion is complete (i.e., whether it has digested 100% of the protein) and the other is whether the digestion process has stopped (i.e., whether it has gone as far as it is going to go, even if it is not 100%). In the second case, a correction factor could be used to determine the actual protein concentration in the original sample, and the values obtained from different samples could still be compared for differential protein expression. In the first case, great care must be taken to minimize the effect of different digestion times. If there are a limited number of target proteins, the digestion protocol should be optimized for the particular protein targets.

That being said, the technical variability (i.e., the variability determined by repeated MRM-MS analysis of the same sample) has been shown to be <6% [5] while the variability for the entire experiment (variously termed analytical or experimental variability) has been shown to be <20% for MRM assays of proteins in undepleted plasma (Fig. 3) [5]. In this study, 45 peptide targets were analyzed 12 times, and the CVs were compared using only the data for the endogenous peptide (blue bars), and using an equimolar mixture of the SIS peptide standards. As can be seen from the figure, there was a dramatic shift in the median CV – from ~25–30% CV to 4–6% CV. The use of a set of SIS peptides whose concentrations were balanced to match those of the endogenous peptide resulted in a further decrease in % CV, to 2–4%.

1.3 Development of an MRM Method

The steps in development of an MRM method have been described in detail in two of our recent publications [4, 5]. Briefly, published libraries of MRM transitions can be used as "starting points," but empirical tuning of the source parameters, such as the collision-energy, can lead to an improvement in the sensitivity of the overall assay. Initially, a set of five transitions per protein is selected – usually peptides with high sensitivities; however, peptides containing variable residues or residues that can be oxidized (including cysteines and methionines) should be avoided. Sufficient data points should be taken across each peak for accurate determination of the peak areas. For the final assay, the best three transitions are used. In addition, the concentrations of the SIS peptides are adjusted to be as close as possible to the expected concentrations of their endogenous counterparts (Fig. 3, right); not surprisingly, the more abundant transitions produce the best CVs.

Newer quantitation software, such as the Agilent MassHunter Quant software, actually bases the quantitation on only the most abundant "quantifier" transition, with the other two transitions (the qualifier transitions) being used to ascertain that

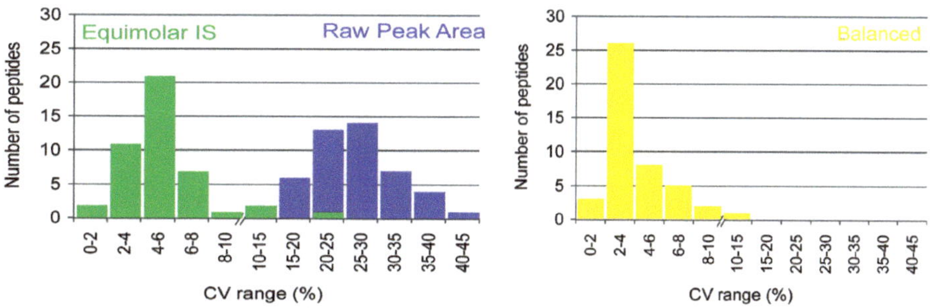

Fig. 3 Reproducibility of 45 targets, 12 replicates. *Blue*: % CVs based on raw peak areas; *green*: % CVs using a set of equimolar internal standards; *yellow*: % CVs using concentration-balanced internal standards. Note the shift towards lower CVs with SIS peptides and a "concentration balanced" mixture of SIS peptides. Adapted from [5], with permission

the peptide monitored is the correct target, i.e., by comparing the elution times, the fragment ion ratios, and the peak profiles (Fig. 4). This validation is carried out both in buffer and in the sample matrix (in this case, undepleted plasma).

There is still the possibility that interference may be encountered in a patient sample that was not present in the pooled plasma sample. For this reason we recommend using more than one proteotypic peptide for quantifying proteins. Table 1 shows the quantitative analysis of alpha-1-antitrypsin using five peptides. The results between peptides differ by a factor of 30 but the results obtained from the individual peptides show a good correlation with correlation coefficients $r > 0.9$ (Fig. 5a), even though the absolute concentration values based on each peptide are different. These differences are probably due to differences in digestion efficiencies for the different peptides, with the highest calculated concentration probably being closest to the actual concentration in the sample. The good correlation, however, indicates that, even for these other peptides, the relative peptide/protein concentrations can still be used even though the calculated protein concentrations may be different. If there are specific peptides that are "outliers," this correlation can also be used to detect interferences that may be present in a single sample (Fig. 5b).

This has led to a change in strategy for our multiplexed analysis. First, the transitions are verified to be interference-free using three transitions per peptide and a pooled plasma sample. Because the % CVs for each transition have been determined to be low, multiple peptides per protein are used in place of additional transitions for the same peptide. Ideally, three peptides per protein would be used for each protein in the final method. Using the dwell times of the Agilent 6490, a maximum of ~270 transitions can "fit" into a 30-min analytical run while keeping the dwell time above 10 ms per transition and the target cycle time below 600 ms. Thus, in our latest project, to achieve the highest possible throughput, we actually used only the quantifier transition, because this transition had previously been determined to be free of interferences in previous analyses of a pooled plasma sample [16].

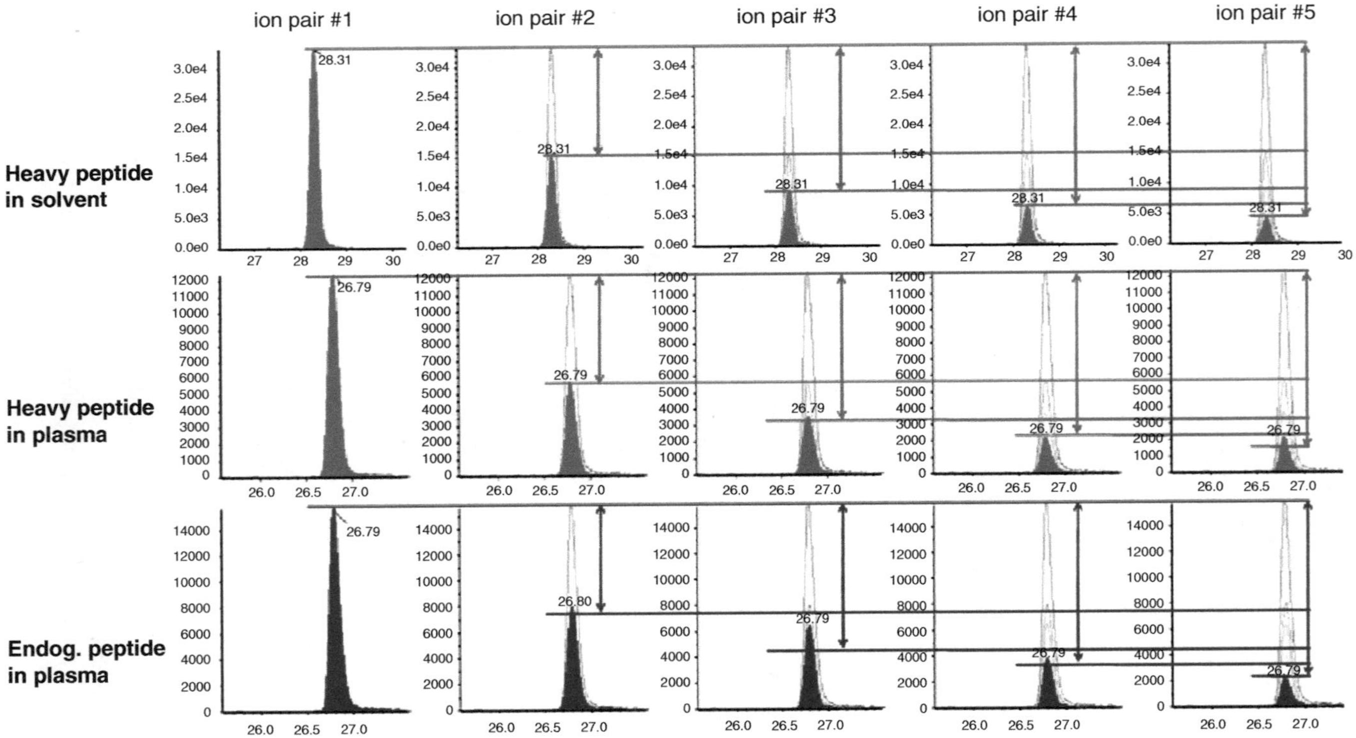

Fig. 4 Determination of interference-free transitions. To check for possible interferences, the retention times and relative peak areas of the SIS peptide in buffer and both the endogenous material and the SIS peptide in a pooled human plasma sample. The retention times of both the SIS and the endogenous peaks in plasma should be the same (although they might be shifted slightly from those in buffer). The ratios of the precursor to the fragment ions, and the fragment ions to each other, for both the SIS and the endogenous peptide in plasma should be the same as corresponding ratios for the SIS peptide in buffer. If they do not, this indicates an interference from the plasma matrix. In this case, there is a significant interference from the plasma background in the endogenous peptide for transition #3, and for the heavy SIS peptide in transition #5

Table 1 Calculated protein concentrations based on the experimentally-derived concentrations of various peptides. Adapted from [12] with permission

Protein	Peptide	LOQ (amol) on column	Analytical precision (% CV)	LOQ (ng/mL)	Experimentally-determined plasma protein concentration (ng/mL), based on the MRM analysis of each individual peptide	Expected plasma concentration (ng/mL) based on the literature
α1-Antitrypsin						1,400,000
	ITPNLAEFAFSLYR	2500	4.2	776	954,881	
	LQHLENELTHDIITK	274	2.6	85	53,242	
	LSITGTYDLK	605	3.2	188	1,520,584	
	SVLGQLGITK	411	4.4	128	929,184	
	VFSNGADLSGVTEEAPLK	18,257	7.9	5,665	376,820	

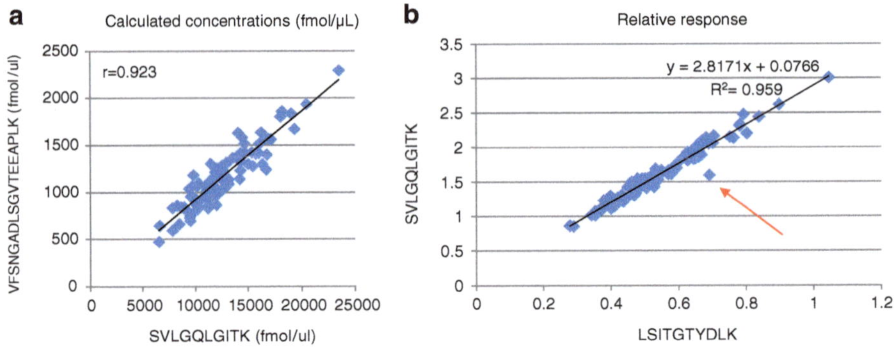

Fig. 5 (**a**) The relative response curve shows a high degree of concentration of observed concentration using different peptides as target. (**b**) An outlier (*arrow*) in the relative response curve indicates an interference with the selected transition in this particular sample. Adapted from [16], with permission

2 Current Research in MRM-Based Clinical Proteomics

2.1 Targeted MRM-Based Quantitation

The major bottleneck in the use of peptide or protein biomarkers in the clinic is the mismatch between the throughput of standard LC-MRM techniques and the number of samples required for biomarker verification and validation, which require the analysis of hundreds to thousands and thousands to tens of thousands of samples, respectively [17, 18]. Methods that have been used successfully for biomarker discovery, such as "MudPIT" (multidimensional protein identification technology [19]) which involve fractionation and separate analysis of each fraction (coupled with label-free quantitation methods, such as spectral counting [20] or ion accounting [21]) are too time-consuming and labor-intensive for large-scale studies. Likewise, differential proteomics techniques such as DIGE [22] and iTRAQ [23, 24] are also successful for biomarker discovery but are too costly and labor-intensive for the large numbers of samples required for the validation step.

There has therefore been a great deal of interest in the development of highly-multiplexed and high-throughput MRM techniques, which could quantitate large numbers of proteins in large numbers of samples with a fairly short analysis time. Multiplexed MRM methods have already appeared in the literature. In 2009, Carr and colleagues demonstrated the targeted MRM analysis for 27 peptides, targeting 9 proteins, including 2 moderate-abundance and 4 low-abundance CVD-related proteins in depleted human plasma [25], using a MudPIT strategy [26]. More recently, in 2011, the Aebersold group demonstrated 6,050 transitions targeting 757 peptides in a yeast-cell lysate, in one 60-min experiment, with attomole amounts injected on-column [27]. The authors also stated that the software was capable of even higher degrees of multiplexing, with the possibility of using 10,000 transitions targeting 1,000 peptides in a single analysis. Our group has recently reported the MRM-based analysis of 67 CVD-related proteins in undepleted plasma [16] and of 63 urinary proteins related to bladder cancer [28]. MRM-based methods for FDA-approved protein biomarkers, such as angiotensin I [29] and thyroglobulin [30], are already in use in the clinic.

To improve the throughput and reduce the cost, our laboratory and others have been working towards increasing the numbers of proteins that can be quantitated *directly* from human plasma, without depletion or enrichment steps. Plasma, with its wide dynamic range of protein concentrations (~10^{10} [31]), is a particularly challenging matrix to work with, but it is a readily-available biofluid that comes in contact with diseased or damaged tissues and thus carries within it protein information on many diseases, including cancer and heart disease. Using Agilent's "ion-funnel" technology, we have recently reported a study where 67 plasma proteins could be quantitated in 30 min using 135 proteotypic peptides [32]. The use of normal flow rates instead of nano-electrospray allowed the use of 2.1 mm-i.d. columns which contain more packing material than nanoscale capillary columns. This improved the robustness of the assay and shortened the analysis time. More sample could be injected into this high-flow system, which, in most cases, was able to overcome the loss of sensitivity encountered when moving away from nano-electrospray, resulting in detection limits with attomole level limits of quantitation, the equivalent of low nanogram/milliliter levels (<20% CV and accuracy 80–120%) for 81 of the 135 peptides. Using this instrument, the highest seven or eight out of the expected ten orders of magnitude of protein concentrations in plasma could be quantitated (Fig. 6) [59].

An extensive cataloging of identified human plasma proteins has recently been performed by Farrah et al. [33], producing a high-confidence non-redundant human plasma proteome reference set with ~2,000 proteins [33]. This has been achieved by compiling tandem MS measurement data from 91 experiments using different sample preparation and MS analysis techniques, and processing it through a uniform data analysis pipeline. Rough estimates of protein concentrations, performed using spectral counting, revealed that ~1,000 proteins should be detectable at >1 ng/mL, with ~600 proteins at a level of >10 ng/mL. This indicates that there is a large group of plasma proteins that are accessible by MS-based analysis but which are not always detected by simple "shotgun" proteomic approaches. In our work we have shown that the latest generation triple quadrupole, combined with high-flow UHPLC separation in MRM mode, allowed the detection and accurate quantitation of plasma proteins down to the 10 ng/mL level in non-depleted, unfractionated plasma samples. In fact, 40% of the MRM assays for these CVD proteins had LOQs between 2 and 100 ng/mL. This indicates that the MRM approach, in combination with SIS peptides and using the latest in MS technology, should be capable of accurately and simultaneously quantitating the top 500–600 plasma proteins in non-depleted, unfractionated plasma.

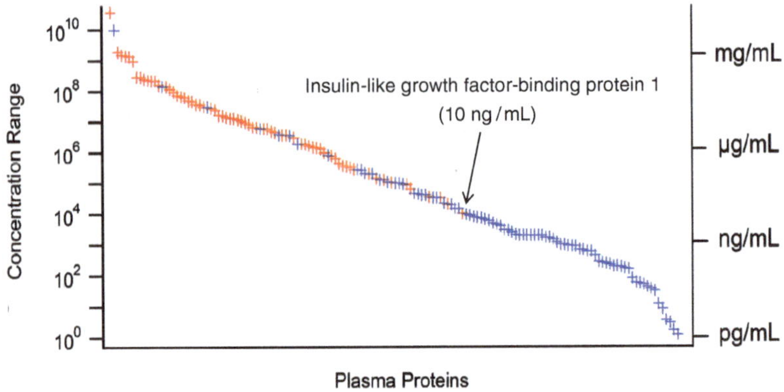

Fig. 6 Concentration range in plasma. *Red* indicates proteins which have already been quantitated in MRM analyses of undepleted plasma

2.2 Multiplexed MRM for Biomarker Discovery

The advantage of a large highly-multiplexed assay is that it can also be used for new biomarker discovery. One advantage of this approach is that the AUC for a *panel* of proteins might show improved diagnostic accuracy over that for a single protein. This was the case in one of our recent projects, which was a study of potential biomarkers of inflammation and anemia [28]. In this study, a panel of 14 peptides representing proteins known or suspected to be involved in iron deficiency was examined in mouse models of iron-deficiency anemia, inflammation, and the combination of anemia and inflammation. A panel of eight biomarker proteins [apolipoprotein A4 (APO4), transferrin, transferrin receptor 1, ceruloplasmin, haptoglobin, lactoferrin, hemopexin, and matrix metalloproteinase-8 (MMP8)] could distinguish between normal mouse plasma and plasma from mice with the different diseases. Within this set of eight proteins, transferrin showed the best classification accuracy for a single protein over all samples (72%) and within the normal group (94%). Compared to the best single-protein biomarker, however, the use of the composite eight-protein biomarker panel improved the classification accuracy from 94% to 100% in the normal group, from 50% to 72% in the inflammation group, from 66% to 96% in the iron-deficiency anemia group, and from 79% to 83% in the inflammation plus iron-deficiency anemia group (Fig. 7).

2.3 MRM for Biomarker Verification

In another recent study, a series of urinary protein biomarkers found in an iTRAQ study was validated using MRM analysis of proteins normally found in plasma [34]. MRM analysis of urine from patients with bladder cancer, urinary tract infection,

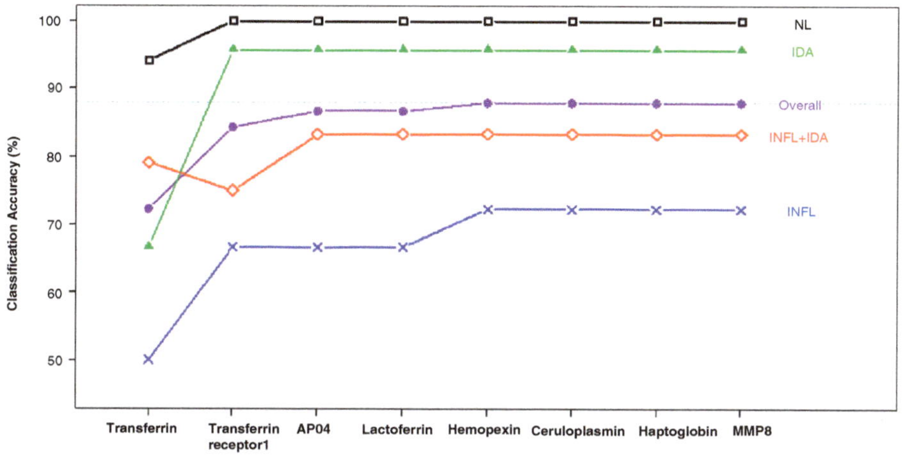

Fig. 7 Improvement in classification accuracy using multiple protein biomarker panels. Classification performance of the multiprotein panel. This figure shows the classification accuracy (y-axis) as additional proteins are incorporated into a multivariate classifier panel (x-axis) by a stepwise discriminant analysis algorithm. *NL* normal, *IDA* iron deficiency anemia, *INFL* inflammation. Adapted from [28], with permission

and hematuria was carried out. Sixty-five peptides representing 63 proteins were studied in 156 patients. The multiplexed MRM-MS data was used to generate a six-peptide marker panel (afamin, adiponectin, complement C4 gamma chain, apolipoprotein A-II precursor, ceruloplasmin, and prothrombin) which could discriminate bladder cancer subjects from patients without cancer with an AUC of 0.814, characterized by a 76.3% positive predictive value, and a 77.5% negative predictive value (Fig. 8).

2.4 *MRM for Pathway Analysis*

One very interesting application of MRM, pioneered by the Koomen laboratory in 2011, has been the development of "libraries" of transitions designed to probe cancer-related pathways [35]. This library has been designed to facilitate quantitative proteomics studies of cancer-related pathways, including signaling pathways, related to colon, lung, melanoma, leukemia, and myeloma, by combining SDS-PAGE with LC-MRM-MS. Thus far, methods are available for detecting 876 peptides from 218 cancer-related proteins, with 95 quantitative assays that include SIS peptides [The Quantitative Assay Database (QUAD), http://proteome.moffitt.org/QUAD/]. These methods are designed for the proteomics analysis of tissue samples, including cell culture and patient biopsies, and hold the potential for allowing the quantitative assessment of different treatment strategies. New standard methods are being added, and standard peptides are being synthesized on a cost-sharing basis.

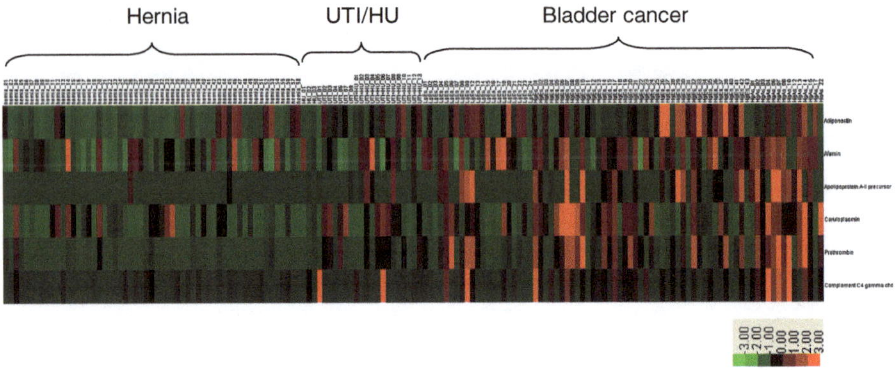

Fig. 8 Heat map of the final 6-protein biomarker panel for bladder cancer. The color represents a standardized value of peptide concentration where the mean of each row (a peptide) is 0. *Red* represents high abundance; *green* represents low abundance. The brightness of the color indicates the number of standard deviations from the average concentration for a specific peptide. Adapted from [34], with permission

2.5 *Stable Isotope Standards and Capture by Anti-peptide Antibodies*

SISCAPA (stable isotope standards and capture by anti-peptide antibodies) is a technique based on the capture of peptides, elution, and analysis by LC-MRM-MS [36]. Because of the affinity capture step, SISCAPA can provide very low detection limits (from the low nanograms per milliliter range down to the low picograms per milliliter range [37], depending on the quantity of plasma used), and recent studies have shown its capability for automation [37, 38]. It may have limited multiplexing capacity, however, because of physical limitations on the number of antibodies, which can be immobilized on the beads. The use of antibodies also results in a higher per-assay cost than methods that do not require antibodies for purification. However, it has recently been demonstrated that the generation of the antibody itself can be five-plexed [39]. In an interesting variation of the SISCAPA protocol, affinity capture of intact proteins instead of peptides (osteopontin splice variants) was used. This was followed by elution, tryptic digestion, dephosphorylation, and LC-MRM-MS quantitation of isomer-specific peptides, using SIS reference standards. Both nanograms per milliliter LODs and good correlation with ELISA results were achieved [40].

2.6 *Matrix-Assisted Laser-Desorption/Ionization-Multiple Reaction Monitoring*

MALDI-MS (not MRM) has recently been described in a paper on the detection of hepcidin in affinity-enriched human plasma [41]. The speed of MALDI-MS acquisition opens up the possibility of performing "standard addition" on each sample. Standard addition is not carried out in normal clinical analysis, but has major advantages in terms of accuracy because individual calibration curves are generated for each sample. At this time, however, as pointed out by the authors, the data analysis was the rate-limiting step, but improvements in software and the increasing use of MALDI in the clinic would make this a high-throughput clinical approach. The Anderson group has also recently promoted the use of SISCAPA with MALDI-MS detection (http://www.genomeweb.com/proteomics/siscapa-based-biomarker-quantitation-moving-maldi-ms-pursuit-higher-throughput-l, 2 December 2011). The immunoMALDI (iMALDI) method [42–45] also uses anti-peptide antibody-based enrichment, but the affinity beads are placed directly on the MALDI target. The analytes are released by the matrix solvent. The iMALDI assay for angiotensin in plasma has demonstrated sensitivities in the low picograms per milliliter range using as little as 10 μL of plasma [44]. Although both of these techniques normally utilize MALDI-MS and not MALDI-MRM, MALDI-MRM could be used and would provide additional specificity.

3 Statistics and Data Analysis

For an assay, a clinically-acceptable coefficient of variation (CV) is usually <20% [46]. In addition, the clinical utility of an assay is characterized by its ability to predict an accurate diagnosis. To determine this accuracy, biostatisticians use a technique called area-under-the-curve (AUC) which compares the true positive rate to the false positive rate of an assay (or the true negative rate to the false negative rate). In a clinical setting, if a patient has the disease and the assay tests positive, this is a true positive. If the patient has the disease and tests negative, this is a false negative. If the patient does not have the disease and tests positive, this is a false positive. If the patient does not have the disease, and tests negative, this is a true negative. Thus there are four possibilities for a test result (true positive, true negative, false positive, false negative). When the true positive rate of an assay (y-axis) is plotted vs its false positive rate (x-axis), the result is the receiver operating characteristic (ROC) curve, apparently named because it was first developed for analyzing radar signals in WWII [47].

Equal true positive and false positive rates would give a point on the diagonal, the value obtained by random guessing. The AUC for the line determined by this equal-probability random guessing is 0.5 (Fig. 9). If the values are above the line, then this classifier is producing better than random predictions (with the highest possible value being 1.0, perfect). By definition, AUC values are always $\geq$0.5. For a biomarker assay, AUC values of 0.8 or higher are considered good, although some tests in use today have lower AUC scores. For example, prostate specific antigen (PSA) tests have an AUC value of only 0.72 for free PSA or 0.53 for total PSA [48], but they are used because they are the best tests currently available.

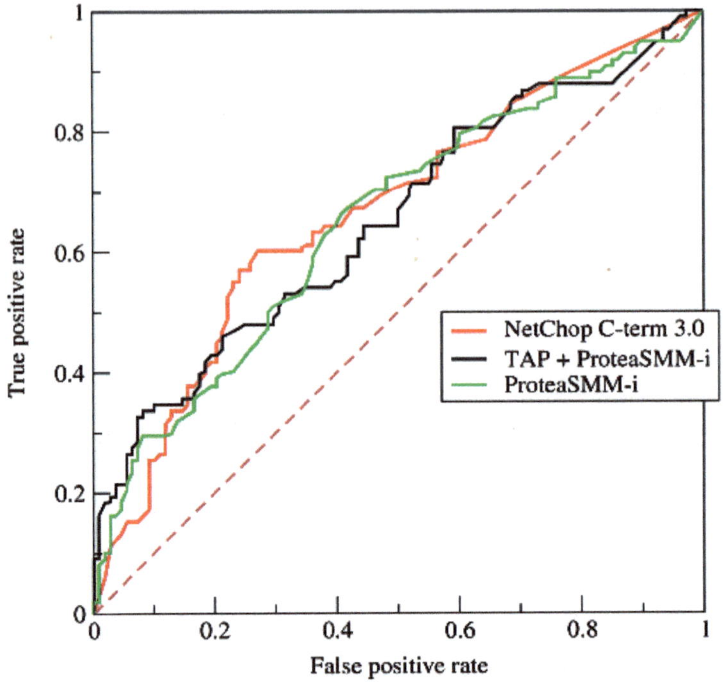

Fig. 9 Examples of ROC curves, with various AUC values (reprinted from [49], with permission)

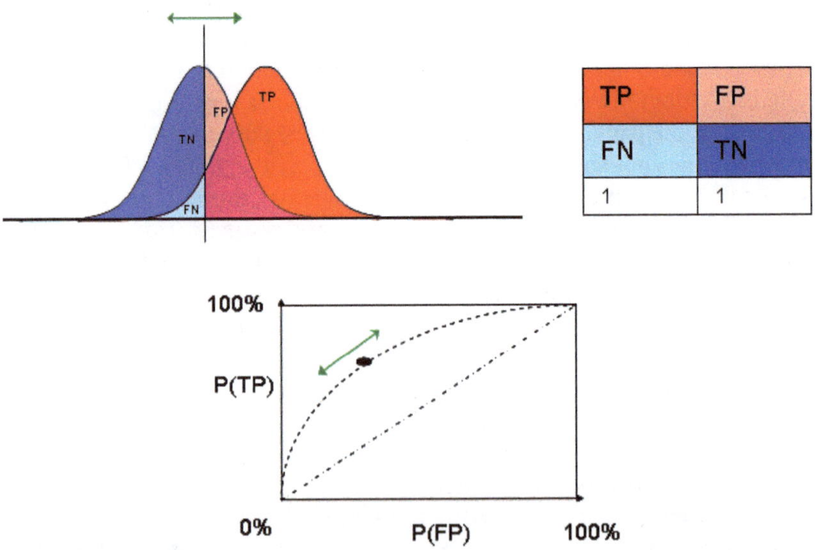

Fig. 10 The effect of the "cut-off" value on the proportion of false positives and negatives (reprinted from [49], with permission)

The tradeoff between specificity and false negatives can easily be seen from the Fig. 10 (adapted from [49]), which assumes that the healthy and disease groups have distributions of biomarker values represented by two partially-overlapping Gaussian distributions. As the "cut-off" line for the diagnostic test is moved to the left or right (as shown by the bi-directional arrow in Fig. 10), the proportion of false positives and false negatives changes.

4 Need for Harmonization

Before a new clinical assay can be widely accepted, it must be compared with other biomarkers using conventional or existing technologies. However, the results from different techniques do not always give the same answers in terms of concentration, as these techniques are subject to different interferences. In a recent study of angiotensin assays, we tried to compare our iMALDI-MS results with ELISA and LC-ESI-MRM analysis. This turned out to be more challenging than expected. First, the commercially-available standard was hygroscopic, which made weighing uncertain, so the actual concentration could have been less than calculated. Second, the standard also contained unknown impurities (Fig. 11a) that had partially correct sequences (Fig. 11b). If the concentration of the standard solutions had been prepared based on weighing out this impure material, it would have led to a "standard" that was too low in the concentration of authentic angiotensin. As a result, because they were measured against this inaccurate standard, the endogenous angiotensin levels as measured by mass spectrometric techniques would have been falsely elevated. Thus the presence of the impurities in the standard would have affected the calibration curve.

Moreover, if this impure material had been used as a standard for the ELISA assay, it is possible that these impurities might have been captured and measured as "authentic" angiotensin I by the anti-angiotensin I antibody in the ELISA assay, because of their partially correct sequences. This could have led to the MS-based results being lower than the results from ELISA In fact, the impurity *was* captured by the anti-angiotensin antibody used in the iMALDI analysis (Fig. 11c) because it contained the VYIH sequence, which is also present in the Ang-I sequence, DR*YVIH*PFHL. However, these impurities would not have been measured in the MS-based assay because they do not have the correct molecular weight. Thus, if this impurity were present in a sample, it would *not* have caused errors in the mass spectrometry-based analyses.

In the case of angiotensin, fortunately there was a "gold standard" pre-weighed standard sample available. This type of standard needs to be available for every single peptide used in the assays determined by multiple techniques and in multiple laboratories.

The need for harmonization in clinical analysis is widely recognized. For example, factors of 5 to >10 have been reported from different methods for

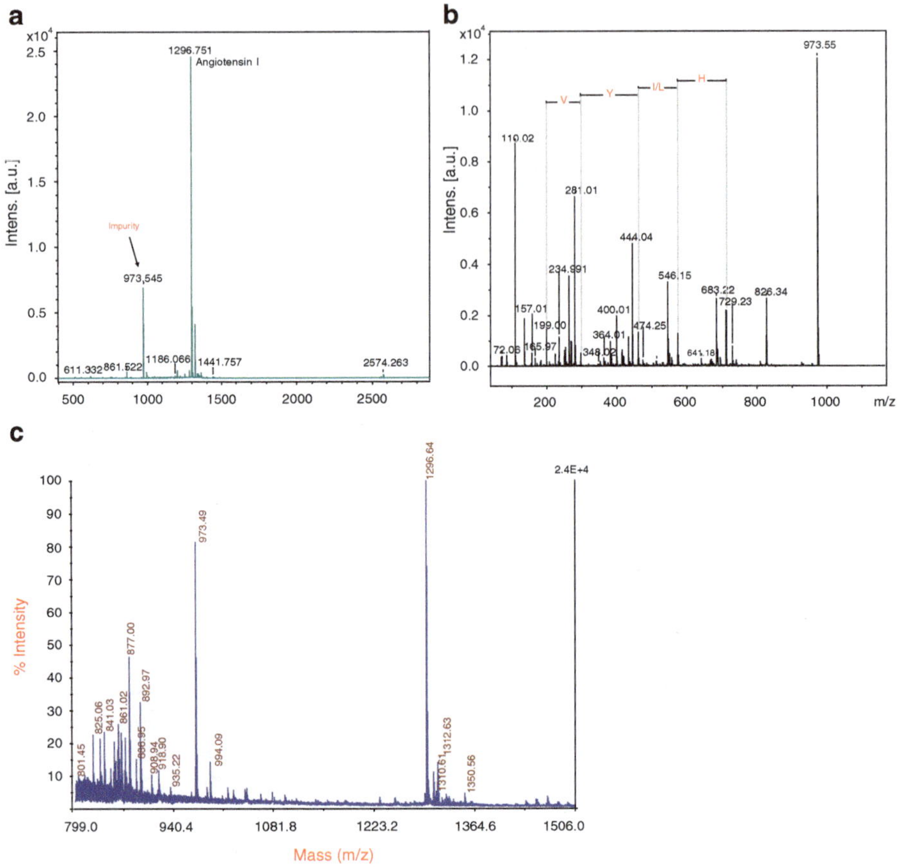

Fig. 11 The effect of impurities in standards. (**a**) MALDI MS of a commercially-available angiotensin I standard; (**b**) MS/MS spectrum of the *m/z* 973 impurity; (**c**) iMALDI experiment showing capture of the impurity as well as the authentic angiotensin standard (*m/z* 1297), by the anti-angiotensin I antibody

determining hepcidin in plasma ([41, 50], respectively), while factors of >150 have been reported for determining hepcidin in urine [50].

5 Need for Standardization

Accurate standards are absolutely necessary for harmonization of the MRM-based results with those of existing techniques, as illustrated by the above example. The results of new mass spectrometry-based results must be accurately compared with those from the older assays. Equally important, the results from new mass spectrometry-based assays must be demonstrated to be reproducible by different laboratories, which might use different instrumentation.

The 2009 report on an inter-laboratory analysis of human plasma using MRM with SIS peptides has demonstrated that the MRM technique is reproducible between laboratories, provided that care is taken with the digestion step and other sample preparation steps [51]. Recently, the Human Proteome Project has recommended the development of anti-peptide antibodies and SIS peptides for every human protein [21, 52, 53]. In addition, the HUPO Standards Initiative has proposed the development of "standard samples" which could be used for instrument and method validation [54–57]. Generating these standards will require the creation of a pooled plasma sample from a defined, large set of individuals that will encompass different health states. This will ensure the representation of normal plasma protein targets, as well as those that might only occur during various acute disease states. The concentration of all of the peptide targets should be defined by a standardized analysis of this reference sample. Users will then be able to assess the analytical performance of their instrumentation and sample preparation protocols by using validated multiplexed MRM assays on this reference sample. These verified standards will also allow protein concentrations to be expressed accurately in terms of nanograms per milliliter, which is what clinicians need.

References

1. Kondrat RW, McClusky GA, Cooks RG (1978) Anal Chem 50:2017
2. Yost RA, Enke CG (1978) J Am Chem Soc 100:2274
3. Harvan DJ, Hass JR, Schroeder JL, Corbett BJ (1981) Anal Chem 53:1755
4. Kuzyk MA, Parker CE, Borchers CH (2012) In: Backvall H (ed) Methods in molecular biology. Humana Press (in press)
5. Kuzyk MA, Smith D, Yang J, Cross TJ, Jackson AM, Hardie DB, Anderson NL, Borchers CH (2009) Mol Cell Proteomics 8:1860
6. Stokvis E, Rosing H, Beijnen JH (2005) Rapid Commun Mass Spectrom 19:401
7. Regnier FE, Riggs L, Zhang RJ, Xiong L, Liu PR, Chakraborty A, Seeley E, Sioma C, Thompson RA (2002) J Mass Spectrom 37:133
8. Zhang R, Sioma CS, Thompson RA, Xiong L, Regnier FE (2002) Anal Chem 74:3662
9. Ong S, Mann M (2005) Nat Chem Biol 1:252
10. Elliott M, Smith D, Kuzyk M, Parker CE, Borchers CH (2009) J Mass Spectrom 44:1637
11. Hulmes JD, Bethea D, Ho K, Huang S-P, Ricci DL, Opiteck GJ, Hefta SA (2004) Clin Proteomics 1:17
12. Aguilar-Mahecha A, Kuzyk MA, Domanski D, Borchers CH, Basik M (2012) PLos One 7: e38290. doi:10.1371journal.pone.0038290
13. Rai AJ, Gelfand CA, Haywood BC, Warunek DJ, Yi J, Schuchard MD, Mehigh RJ, Cockrill SL, Scott GBI, Tammen H, Schulz-Knappe P, Speicher DW, Vitzthum F, Haab BB, Siest G, Chan DW (2005) Proteomics 5:3262
14. Tuck MK, Chan DW, Chia D, Godwin AK, Grizzle WE, Krueger KE, Rom W, Sanda M, Sorbara L, Stass S, Wang W, Brenner DE (2009) J Proteome Res 8:113
15. Proc JL, Kuzyk MA, Hardie DB, Yang J, Smith DS, Jackson AM, Parker CE, Borchers CH (2010) J Proteome Res 9:5422

16. Domanski D, Percy AJ, Yang J, Chambers AG, Hill JS, Cohen Freue GV, Borchers CH (2012) Proteomics 12:1222
17. Paulovich AG, Whiteaker JR, Hoofnagle AN, Wang P (2008) Proteomics Clin Appl 2:1386
18. Surinova S, Schiess R, Hüttenhain R, Cerciello F, Wollscheid B, Aebersold R (2011) J Proteome Res 10:5
19. Wolters DA, Washburn MP, Yates JR III (2001) Anal Chem 73:5683
20. Asara JM, Christofk HR, Freimark LM, Cantley LC (2008) Proteomics 8:994
21. Silva JC, Gorenstein MV, Li G-Z, Vissers JPC, Geromanos SJ (2006) Mol Cell Proteomics 5:144
22. GE_Healthcare (2011) About 2-D fluorescence difference gel electrophoresis (2-D DIGE). http://www.gelifesciences.com/aptrix/upp01077.nsf/content/2d_electrophoresis~new_to_2d_dige. Accessed December 2011
23. Ross PL, Huang YN, Marchese JN, Williamson B, Parker K, Hattan S, Khainovski N, Pillai S, Dey S, Daniels S, Purkayastha S, Juhasz P, Martin S, Bartlet-Jones M, He F, Jacobson A, Pappin DJ (2004) Mol Cell Proteomics 3:1154
24. Cohen Freue GV, Sasaki M, Meredith A, Günther OP, Bergman A, Takhar M, Mui A, Balshaw RF, Ng RT, Opushneva N, Hollander Z, Li G, Borchers CH, Wilson-McManus J, McManus BM, Keown PA, McMaster WR (2010) Mol Cell Proteomics 9:1954
25. Keshishian H, Addona T, Burgess M, Mani DR, Shi X, Kuhn E, Sabatine MS, Gerszten RE, Carr SA (2009) Mol Cell Proteomics 8:2339
26. Washburn MP, Wolters D, Yates JR 3rd (2001) Nat Biotechnol 19:242
27. Kiyonami R, Schoen A, Prakash A, Peterman S, Zabrouskov V, Picotti P, Aebersold R, Huhmer A, Domon B (2011) Mol Cell Proteomics 10:M110.002931
28. Domanski D, Cohen Freue G, Sojo L, Kuzyk MA, Parker CE, Goldberg YP, Borchers CH (2011) J Proteomics. doi:10.1016/j.jprot.2011.11.022
29. Bystrom CE, Salameh W, Reitz R, Clarke NJ (2010) Clin Chem 56:1561
30. Hoofnagle AN, Becker JO, Wener MH, Heinecke JW (2008) Clin Chem 54:1796
31. Anderson NL, Anderson NG (2002) Mol Cell Proteomics 1:845
32. Domanski D, Smith DS, Miller CA, Yang Y, Jackson AM, Cohen Freue G, Hill JS, Parker CE, Borchers CH (2011) Clin Lab Med 31:371
33. Farrah T, Deutsch EW, Campbell DS, Sun Z, Bletz JA, Mallick P, Katz JE, Malmström J, Ossola R, Watts JD, Lin B, Zhang H, Moritz RL, Aebersold R (2011) Mol Cell Proteomics 10. doi:10.1074/mcp.M110.006353
34. Chen Y-T, Chen H-W, Domanski D, Smith DS, Liang K-H, Wu C-C, Chen C-L, Chung T, Chen M-C, Chang Y-S, Parker CE, Borchers CH, Yu J-S (2012) Proteomics. doi:10.1016/j.jprot.2011.12.031
35. Remily-Wood ER, Liu RZ, Xiang Y, Chen Y, Thomas CE, Rajyaguru N, Kaufman LM, Ochoa JE, Hazlehurst L, Pinilla-Ibarz J, Lancet J, Zhang G, Haura E, Shibata D, Yeatman T, Smalley KSM, Dalton WS, Huang E, Scott E, Bloom GC, Eschrich SA, Koomen JM (2011) Proteomics Clin Appl. doi:10.1002/prca.201000115
36. Anderson NL, Anderson NG, Haines LR, Hardie DB, Olafson RW, Pearson TW (2004) J Proteome Res 3:235
37. Whiteaker JR, Zhao L, Anderson L, Paulovich AG (2010) Mol Cell Proteomics 9:184
38. Anderson NL, Jackson A, Smith D, Hardie D, Borchers C, Pearson TW (2009) Mol Cell Proteomics 8:995
39. Whiteaker JR, Zhao L, Abbatiello SE, Burgess M, Kuhn E, Lin C, Pope M, Razavi M, Anderson NL, Pearson TW, Carr SA, Paulovich AG (2011) Mol Cell Proteomics 10: M110.005645
40. Wu J, Pungaliya P, Kraynov E, Bates B (2012) Biomarkers 17:125
41. Anderson DS, Kirchner M, Kellogg M, Kalish LA, Jeong JY, Vanasse G, Berliner N, Fleming MD, Steen H (2011) Anal Chem 83:8357
42. Jiang J, Parker CE, Hoadley KA, Perou CM, Boysen G, Borchers CH (2007) Proteomics Clin Appl 1:1651

43. Jiang J, Parker CE, Fuller JR, Kawula TH, Borchers CH (2007) Anal Chim Acta 605:70
44. Reid JD, Holmes DT, Mason DR, Shah B, Borchers CH (2010) J Am Soc Mass Spectrom 21:1680
45. Shah B, Reid JD, Kuzyk MJ, Parker CE, Borchers CH (2012) Methods in molecular biology (in press)
46. Reed GF, Lynn F, Mead BD (2002) Clin Diagn Lab Immunol 9:1235
47. Green DM, Swets JM (1966) Signal detection theory and psychophysics. Wiley, New York
48. Shariat SF, Semjonow A, Lilja H, Savage C, Vickers AJ, Bjartell A (2011) Acta Oncol 50:61
49. Receiver Operating Characteristic (2011) http://en.wikipedia.org/wiki/Receiver_operating_characteristic. Accessed December 2011
50. Kroot JJC, Kemna EHJM, Bansal SS, Busbridge M, Campostrini N, Girelli D, Hider RC, Koliaraki V, Mamalaki A, Olbina G, Tomosugi N, Tselepis C, Ward DG, Ganz T, Hendriks JCM, Swinkels DW (2009) Haematologica 94:1748
51. Addona TA, Abbatiello SE, Schilling B, Skates SJ, Mani DR, Bunk DM, Spiegelman CH, Zimmerman LJ, Ham A-JL, Keshishian H, Hall SC, Allen S, Blackman RK, Borchers CH, Buck C, Cardasis HL, Cusack MP, Dodder NG, Gibson BW, Held JM, Hiltke T, Jackson A, Johansen EB, Kinsinger CR, Li J, Mesri M, Neubert TA, Niles RK, Pulsipher TC, Ransohoff D, Rodriguez H, Rudnick PA, Smith D, Tabb DL, Tegeler TJ, Variyath AM, Vega-Montoto LJ, Wahlander A, Waldemarson S, Wang M, Whiteaker JR, Zhao L, Anderson NL, Fisher SJ, Liebler DC, Paulovich AG, Regnier FE, Tempst P, Carr SA (2009) Nat Biotechnol 27:633
52. HUPO (2011) The Human Proteome Project. http://www.hupo.org/research/hpp/. Accessed December 2011
53. Anderson NL, Anderson NG, Pearson TW, Borchers CH, Paulovich AG, Patterson SD, Gillette M, Aebersold R, Carr SA (2009) Mol Cell Proteomics 8:883
54. HUPO (2010) The HUPO standards initiative. http://www.hupo.org/research/psi/. Accessed December 2011
55. HUPO (2011) The HUPO proteomics standards initiative. Accessed December 2011
56. Legrain P, Aebersold R, Archakov A, Bairoch A, Bala K, Beretta L, Bergeron J, Borchers C, Corthals GL, Costello CE, Deutsch EW, Domon B, Hanock W, He F, Hochstrasser D, Marko-Varga G, Salekdeh GH, Sechi S, Snyder M, Srivastava S, Uhlen M, Hu CH, Yamamoto T, Paik YK, Omenn GS (2011) Mol Cell Proteomics 10:M111.00993
57. Omenn GS (2004) Proteomics 4:1235
58. Kinter M, Sherman NE (2005) Protein sequencing and identification using tandem mass spectrometry, John Wiley & Sons, USA, p. 58.
59. Percy AJ, Chambers AG, Yang J, Domanski D, Borchers CH (2012) Anal Bioanal Chem doi:10.1007/s00216-00012-06010-y

Top Curr Chem (2014) 336: 139–176
DOI: 10.1007/128_2012_413

Published online: 7 April 2013

Advances in MALDI Mass Spectrometry in Clinical Diagnostic Applications

Eddy W.Y. Ng, Melody Y.M. Wong, and Terence C.W. Poon

Abstract The concept of matrix-assisted laser desorption/ionization mass spectrometry (MALDI MS) was first reported in 1985. Since then, MALDI MS technologies have been evolving, and successfully used in genome, proteome, metabolome, and clinical diagnostic research. These technologies are high-throughput and sensitive. Emerging evidence has shown that they are not only useful in qualitative and quantitative analyses of proteins, but also of other types of biomolecules, such as DNA, glycans, and metabolites. Recently, parallel fragmentation monitoring (PFM), which is a method comparable to selected reaction monitoring, has been reported. This highlights the potentials of MALDI-TOF/TOF tandem MS in quantification of metabolites. Here we critically review the applications of the major MALDI MS technologies, including MALDI-TOF MS, MALDI-TOF/TOF MS, SALDI-TOF MS, MALDI-QqQ MS, and SELDI-TOF MS, to the discovery and quantification of disease biomarkers in biological specimens, especially those in plasma/serum specimens. Using SELDI-TOF MS as an example, the presence of systemic bias in biomarker discovery studies employing MALDI-TOF MS and its possible solutions are also discussed in this chapter. The concepts of MALDI, SALDI, SELDI, and PFM are complementary to each other. Theoretically, all these technologies can be combined, leading to the next generation of the MALDI MS technologies. Real applications of MALDI MS technologies in clinical diagnostics should be forthcoming.

E.W.Y. Ng and T.C.W. Poon (✉)
Department of Paediatrics, The Chinese University of Hong Kong, Prince of Wales Hospital, Shatin, N.T., Hong Kong SAR, China

Li Ka Shing Institute of Health Sciences, The Chinese University of Hong Kong, Prince of Wales Hospital, Shatin, N.T., Hong Kong SAR, China
e-mail: tcwpoon@cuhk.edu.hk

M.Y.M. Wong
Li Ka Shing Institute of Health Sciences, The Chinese University of Hong Kong, Prince of Wales Hospital, Shatin, N.T., Hong Kong SAR, China

Keywords Biomarkers · Clinical diagnostics · Matrix-assisted laser desorption/ionization (MALDI) · Surface assisted laser desorption/ionization (SALDI) · Surface-enhanced laser desorption/ionization (SELDI) · Time-of-flight (TOF)

Contents

1 Introduction 141
2 MALDI-TOF MS and MALDI-TOF/TOF MS 142
2.1 Basic Principles 142
2.2 A High-Throughput Technology for Discovery and Quantification of Biomarkers 144
2.3 Common Use in Analyses of Large Biomolecules, But Not Small Biomolecules 144
2.4 Quantitative Issues in the MALDI-TOF Mass Spectra 145
2.5 Coupled with Functionalized Magnetic Beads for Peptide/Protein Biomarker Discovery 145
2.6 Sequence-Specific Exopeptidase Activity Test for "Functional" Biomarkers in Disease Diagnosis 147
2.7 Quantification of Protein Biomarkers in Disease Diagnosis 147
2.8 Identification of Disease Associated Aberrant Glycosylation 149
2.9 Qualitative and Quantitative Analysis of Genetic Markers 151
2.10 Quantification of Metabolites by MALDI-TOF MS and SALDI-TOF MS 152
2.11 Discovery of Metabolite Biomarkers by Quantitative Profiling 153
2.12 Quantification of Metabolites by MALDI-TOF/TOF MS 154
3 MALDI-QqQ MS 156
3.1 Quantification of Biomarkers 156
4 SELDI-TOF MS 157
4.1 Basic Principle 157
4.2 Quantitative Issues in the SELDI-TOF Mass Spectra 158
4.3 General Biomedical Applications of SELDI-TOF MS 159
4.4 High-Throughput Technology for Biomarker Discovery 159
4.5 Host Response Proteins Forming the SELDI Proteomic Fingerprints 161
4.6 Presence of Systemic Bias in Biomarker Discovery Studies 161
4.7 Overcoming Systemic Bias in Biomarker Discovery Studies 162
5 Future Prospectives 165
References 166

Abbreviations

2D-PAGE	Two-dimensional polyacrylamide gel electrophoresis
AFP	Alpha-fetoprotein
CDG	Congenital disorders of glycosylation
CHCA	Cyano-4-hydroxycinnamic acid
CV	Coefficient of variation
DHB	Dihydroxybenzoic acid
EGFR	Epidermal growth factor receptor
ESI	Electrospray ionization
FTICR	Fourier transform ion cyclotron resonance
HbA1c	Glycohemoglobin
HBV	Hepatitis B virus

HCC	Hepatocellular carcinoma
hCG	Human chorionic gonadotropin
IMAC	Immobilized metal affinity chromatography
iMALDI	Immuno-matrix-assisted laser desorption/ionization
IVDMIA	In vitro diagnostic multivariate index assay
LC	Liquid chromatography
m/z	Mass-to-charge
MALDI	Matrix-assisted laser desorption/ionization
MRM	Multiple reaction monitoring
MS/MS	Tandem mass spectrometry
PFM	Parallel fragmentation monitoring
Pro-apoC2	Proapolipoprotein CII
QqQ	Triple quadrupole
SAA	Serum amyloid A
SALDI	Surface-assisted laser desorption/ionization
SARS	Severe acute respiratory syndrome
SELDI	Surface-enhanced laser desorption/ionization
SISCAPA	Stable isotope standards and capture by anti-peptide antibodies
SNP	Single nucleotide polymorphism
SRM	Selected reaction monitoring
SSEAT	Sequence-specific exopeptidase activity test
TOF	Time-of-flight

1 Introduction

The history of matrix-assisted laser desorption/ionization (MALDI) mass spectrometry (MS) can be dated back to 1985 [1]. Karas et al. first reported the use of an organic molecule as a matrix to assist desorption/ionization of other small molecules under UV laser irradiation [1]. In 1987, Koichi Tanaka and his colleagues showed that coupling MALDI to a time-of-flight (TOF) mass analyzer allowed the detection of macromolecules, especially proteins [2]. Koichi Tanaka's MALDI-TOF MS method for analyses of macromolecules was highly regarded. It created new opportunities for application of MS to biomedical research. In 2002, Koichi Tanaka together with two other chemists were awarded The Nobel Prize in Chemistry 2002 for their developments of soft desorption ionization methods for mass spectrometric analysis of biological macromolecules. In the past 15 years, while applications of MALDI MS technologies to qualitative and quantitative analyses of proteins and metabolites have been investigated, these technologies have been widely employed in proteomic/metabolomic research, especially in biomarker discovery. Because this chapter is aimed at updating readers on the advances in MALDI MS technologies in clinical diagnostic applications, it is not going to provide a comprehensive review on all MALDI MS technologies. This chapter will only cover MALDI-TOF MS, MALDI-TOF/TOF MS, SALDI-TOF MS, MALDI-QqQ MS, and SELDI-TOF MS, which have great potentials in influencing the clinical diagnostic practices. Basic principles and brief overviews of these technologies will be provided to an extent allowing readers to understand the potentials and limitations of these technologies in diagnostic applications. Then the applications of these technologies to biomarker discovery and their potential uses in biomarker quantification will be reviewed. Practical concerns and their possible solutions on applying MALDI-based MS technologies, especially SELDI, to quantification and discovery of serum/plasma biomarkers will also be addressed.

2 MALDI-TOF MS and MALDI-TOF/TOF MS

2.1 *Basic Principles*

MALDI is regarded as a soft desorption ionization method because it can result in the formation of ions without significantly breaking any chemical bonds by using optimal laser irradiance [1]. This is particularly important in obtaining the correct mass of a biomolecule, especially for proteins, during MS analysis. Subsequently, structure or sequence information can be obtained by tandem MS analysis. As indicated from the name MALDI, a matrix is needed to assist desorption and ionization of an organic molecule under UV irradiation [1, 3]. Desorption/ionization efficiencies of different types of biomolecules depend on the chemical used as the matrix. For example, cyano-4-hydroxycinnamic acid (CHCA) is a very good matrix for peptides [4]. Sinapinic acid is good for intact proteins [4]. Super-DHB (i.e., a mixture of 2,5-dihydroxybenzoic acid and 2-hydroxy-5-methoxybenzoic acid) is good for glycan analysis [5]. 3-Hydroxypicolinic acid is commonly used for MALDI-TOF MS analysis of DNA molecules [6]. An analyte or a mixture of analytes needs to mix with a chemical matrix in solution phase, and is added on a conductive MS sample plate. After drying, analyte-matrix co-crystals are formed. This work-flow is illustrated in Fig. 1. They are then subjected to MALDI MS analysis. Under UV irradiation the analytes will be desorbed and ionized [3]. In the absence of alkali metal ion and/or halide ion in the co-crystals, singly charged protonated and deprotonated molecules are usually formed. The mass (or molecular weight) of a molecule will be approximately equal to "m/z value -1.0073" and "m/z value $+1.0073$", respectively. In the presence of alkali metal ion and/or halide ion, such as Na^+ and Cl^-, metal ion adducts and/or halide ion adducts may be formed. Either positively or negatively charged molecules are transferred to the mass analyzer for separation according to their mass-to-charge (m/z) ratios.

There are various types of mass analyzers. A TOF or TOF/TOF mass analyzer is the most commonly used coupled with a MALDI source. When kinetic energy is

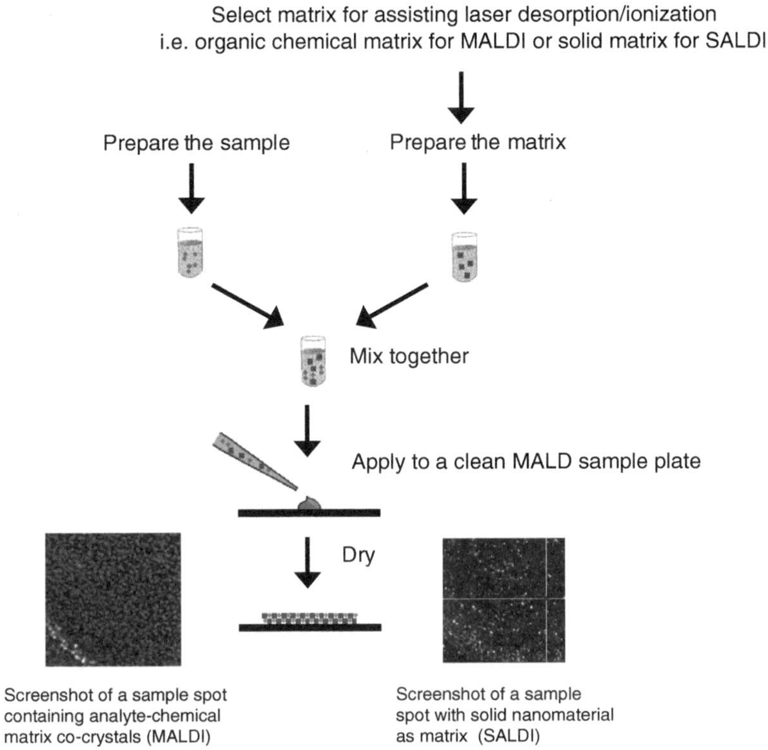

Fig. 1 A typical workflow of sample spot preparations for analyses by MALDI-TOF MS and SALDI-TOF MS. In conventional MALDI-TOF MS, the analytes and chemical matrix are mixed and dried to form co-crystals, whereas analytes are coated homogeneously and distributed evenly on a layer of solid matrix in SALDI-TOF MS

given to a group of charged molecules in direct proportion to their charge states under vacuum, the charged molecules will travel in a flight tube at a velocity inversely proportional to the square root of their m/z values [7]. In other words, charged molecules with larger m/z values have longer TOF, and they are efficiently separated for generating a mass spectrum within 1 s. The resolving power of a TOF mass analyzer depends on the length of the flight path. In contrast, other commonly used mass analyzers, such as ion trap, orbitrap, and Fourier transform ion cyclotron resonance (FTICR) mass analyzers, have resolving powers directly proportional to the time of the charged molecules staying inside the mass analyzers. The high-throughput nature of a TOF mass analyzer makes it perfectly match with a MALDI source. A single TOF mass analyzer does not allow efficient structure/sequence elucidation of a targeted analyte in a mixture of analytes. This can be overcome by linking two TOF mass analyzers in series, i.e., TOF/TOF. The first TOF mass analyzer is used to resolve and select the precursor ion of a targeted analyte for later fragmentation, whereas the second TOF mass analyzer is used to separate the fragment ions for generating a tandem MS spectrum [8, 9].

2.2 A High-Throughput Technology for Discovery and Quantification of Biomarkers

Discovery of disease-specific biomarkers for assisting diagnoses is still a difficult but important task all over the world. One commonly used approach for identification of disease-specific biomarkers is to compare the quantitative biomolecule profiles of plasma/serum specimens from the patients with the target disease and control subjects without the disease. Because of high heterogeneity in the baseline concentrations of various circulating biomolecules among both the patients and control subjects, it is important to obtain and compare quantitative plasma/serum biomolecule profiles from patient and control groups with a reasonable sample size. In such a case, high-throughput technologies are required in order to complete the analyses of the specimens within an acceptable period of time.

The major advantage of MALDI-TOF MS is that it is high-throughput in nature. After preprocessing, samples for MS analyses are applied on a MALDI sample plate as individual spots. MALDI-TOF MS analysis of each sample spot takes less than 1 min. One hundred samples can be automatically analyzed within 1 h. In contrast to electrospray ionization (ESI), which is another commonly used soft ionization technology, a single preprocessed sample is usually subjected to liquid chromatography (LC) before ESI MS analysis [10], resulting in a turn-around time of 20 min to 1 h. Therefore, it takes 30–100 h for analyzing 100 samples by ESI MS. For a shotgun proteomic profiling approach, it will take a day for obtaining a coarse proteomic profile of a single specimen, or at least a week for obtaining a comprehensive proteomic profile.

2.3 Common Use in Analyses of Large Biomolecules, But Not Small Biomolecules

It is well known that MALDI-TOF MS and MALDI-TOF/TOF MS have been widely applied to protein identification in proteomics laboratories. When subjected to MALDI-TOF MS, peptides and proteins are predominantly detected as singly charged protonated molecules at high sensitivities. On one hand, the detection sensitivity of MALDI-TOF MS depends on the chemical composition of a molecule. On the other hand, in general the detection sensitivity is inversely proportional to the mass of a molecule. For large proteins, e.g., albumin, usually at least an amount of 100 fmol to 1 pmol is required for a reliable MS signal. For a clean preparation, a peptide of 0.25 fmol can be readily detected. MALDI-TOF MS can efficiently obtain the masses of majority of peptides in a protein tryptic digest in the MS range of m/z 1,000–2,500 with high accuracy (<40 ppm for external m/z calibration; <5 ppm for internal m/z calibration). The resulted list of tryptic peptides' peak intensities and masses can then be subjected to a database search to obtain the protein identity by using the tryptic peptide mass fingerprinting algorithms, e.g., Mascot [11]. For individual tryptic peptides, it can be further subjected to tandem MS to obtain a series of a-, b-, y-ions if one has a MALDI-TOF/TOF MS instrument. The resulting list of fragment ions' peak intensities and masses can be further subjected to a database search to obtain the protein identity by using the MS/MS ion search algorithms [12].

As described in the previous section, analytes are embedded in analyte-matrix co-crystals before MS analysis. The majority of the matrices are derivatives of benzoic acid, cinnamic acid, and carboxylic acids [3]. However, in MALDI-TOF MS, these small molecules themselves form protonated ions, fragment ions, and cluster ions, and cause intensive chemical noises in the mass range below *m/z* 800 [13, 14]. These noises cause significant interference during the analyses of small molecules. This explains why there have been only a few reports on using MALDI-TOF MS for small molecule analysis in the past 27 years [14]. Although analyses of small molecules are technically difficult, all these reports have provided concrete evidence that MALDI-TOF MS is a feasible tool for analysis of small biomolecules, including amino acids [15], lipids [16], O-linked glycans [17], steroid hormone [18], etc.

2.4 *Quantitative Issues in the MALDI-TOF Mass Spectra*

In addition to matrix chemical noises in the low mass region, the prerequisite formation of analyte-matrix co-crystals causes uneven distribution of analytes on a sample spot. MS signals of various analytes vary significantly among the co-crystals. In common practice a representative mass spectrum of a single sample spot is generated from the summation of mass spectra obtained at different positions within a sample spot. As a result, conventional MALDI-TOF MS methods are usually considered not quantitative, or at most semi-quantitative [4]. However, the reproducibility of the peak intensities of biomolecules in a MALDI-TOF MS spectrum can be improved by using an unbiased automatic mass spectrum acquisition protocol across a sample spot and by providing a fine network (e.g., nitrocellulose coating) for formation of a layer of homogeneous small analyte-matrix co-crystals [4]. With an unbiased MS acquisition protocol and the use of nitrocellulose film, intra-assay and inter-assay coefficients of variation (CVs) of the normalized peak intensities of peptide/protein standards were found to be $<15\%$ [4], suggesting that MALDI-TOF MS is a feasible tool for profiling and quantifying peptides and proteins in biological samples.

2.5 *Coupled with Functionalized Magnetic Beads for Peptide/ Protein Biomarker Discovery*

Plasma/serum samples are highly complex, and cannot be directly subjected to MALDI-TOF MS analysis because of a signal suppression problem. This can be solved by using chromatographic techniques to enrich a subgroup of proteins with matched physicochemical properties. This concept was first introduced by Bruker Daltonics Inc. (Bremen, Germany) as a commercially available system called ClinProt for semi-quantitative profiling of proteins/peptides in serum/plasma. In this system, various types of functionalized magnetic beads with different chromatographic properties are available, including hydrophobic interaction (C3, C8, C18), weak cation exchange, weak anion exchange, metal ion affinity (Cu^{2+}, Fe^{3+}), and lectin affinity (Concanavalin A). The ClinProt magnetic bead technology is only licensed to be performed with MALDI-TOF MS instruments from the same manufacturer. The ClinProt system users are supplied with a kit of standard protocol, together with specific buffers. The compositions of the binding and washing reagents are not disclosed. Because MALDI-TOF MS is a sensitive technology for detection of proteins, only 5 μL of plasma/serum is required for the ClinProt system, according to the supplier's instructions. The eluted proteins/peptides are added on a thin-layer of CHCA, and subjected to MALDI-TOF MS for obtaining a semi-quantitative mass spectrum. The ClinProt technology was first reported to be highly quantitative. The CVs for the normalized protein/peptide peak intensities were $\leq$7% [19]. However, later studies showed that the CVs were between 20% and 30% for both manual and robotic assays [20, 21]. There are about 40 reports on using the ClinProt system in discovery of potential biomarkers of human diseases, such as oral cancer [19], head and neck cancer [20], and nephrotic syndrome [22].

In 2007, Jimenez et al. reported an automated method comparable to the ClinProt system by using C18 hydrophobic magnetic beads for profiling of serum peptides with masses in the range of m/z 800–4,000 [23]. The intra-assay and inter-assay CVs were 2–38% and 10–53%, respectively. Later our group developed a strategy for quantitative profiling of both serum peptides/proteins and micropreparative purification of the corresponding peptides/proteins in parallel using C18 hydrophobic, strong anion exchange, and weak cation exchange magnetic beads [24]. In our method, only 2 μL of serum is required, and sinapinic acid is used as the chemical matrix. By using an automatic platform for the binding, washing, and elution steps and using a MALDI-TOF MS instrument optimized for quantitative proteomic profiling, both intra-assay and inter-assay CVs were found to be 4–30%. Because the peptides/proteins corresponding to the potential diagnostic peaks are purified in parallel with the profiling experiments, the subsequent work for deciphering protein identities of the potential biomarker peaks is greatly simplified. Using this method, we have recently identified proapolipoprotein CII (Pro-apoC2) and a des-arginine variant of serum amyloid A (SAA) as host response biomarkers for diagnosis of late-onset septicemia and necrotizing enterocolitis in preterm infants [25]. The ApoSAA score computed from plasma apoC2 and SAA concentrations was effective in identifying necrotizing enterocolitis/late-onset sepsis cases in both independent case–control and prospective cohort studies. On the basis of the ApoSAA score, infants suspected with the diseases could be stratified into different risk categories. This enabled neonatologists to withhold treatment in 45% and enact early stoppage of antibiotics in 16% of non-sepsis infants.

2.6 *Sequence-Specific Exopeptidase Activity Test for "Functional" Biomarkers in Disease Diagnosis*

The combined use of hydrophobic magnetic beads and MALDI-TOF/TOF MS allows both quantitative profiling of plasma/serum peptides and direct identification of the amino acid sequences of the peptides without the need for subsequent purification work. When Villanueva et al. attempted to identify the serum peptide pattern associated with metastatic thyroid cancer by undertaking this approach, they found that the majority of the disease-associated peptides were derived from fibrinopeptide A, complement C3f, and fibrinogen-α as a result of exopeptidase degradation [26]. It was speculated that proteases produced by the thyroid cancer cells led to the formation of these disease-associated peptides [27]. This led them to develop further the Sequence-Specific Exopeptidase Activity Test (SSEAT) test [27]. Instead of identification of the disease-associated peptides, the test monitors degradation of artificial substrates in the presence of individual patients' sera by MALDI-TOF MS. Double labeled, non-degradable peptides are spiked into the samples as internal standards at the same time to adjust for the adsorptive and processing-related losses. The peak intensity ratios of degradation products to the corresponding non-degradable reference peptides are used as biomarkers. The CVs of these ratios were reported to be 6.3–14.3%. Using the SSEAT test, the group could classify 48 metastatic thyroid cancer patients and 48 healthy controls at 94% sensitivity and 90% specificity [27]. The major advantage of the SSEAT test is that reproducibility problems related to sample collection, storage, and handling in serum peptide profiling analysis can be greatly reduced. Furthermore, theoretically, by using specific peptide sequences as substrates for different diseases, the diagnostic sensitivity and specificity of a SSEAT test may be further improved.

2.7 *Quantification of Protein Biomarkers in Disease Diagnosis*

Immunosorbent assay and immunoturbidity assay are the most commonly used technologies for quantification of specific protein biomarkers in routine clinical chemistry laboratories. Both technologies require the use of specific antibodies. The use of antibodies allows sensitive and specific quantification of a target protein biomarker. However, the use of antibodies can cause uncertainty in measurement. Affinity and specificity of the antibody preparations against a specific antigen varies significantly from source to source. It is not uncommon for immunoassay kits from different manufacturers to produce disconcordant readings. Furthermore, a specific protein can appear as different forms in biological specimens, including glycosylation variants, free subunits, and metabolized forms. A typical example is circulating human chorionic gonadotropin (hCG), which is a useful biomarker for diagnosis of pregnancy, hydatidiform mole, and certain poorly differentiated cancers. HCG is present in a number of forms in blood, including intact hCG, nicked hCG, hyper-

and hypoglycosylated hCG, hCG missing the C-terminal extension, free alpha-subunit, large free alpha-subunit, free beta-subunit, nicked free beta-subunit, and beta-core fragment [28]. For blood samples collected in normal pregnancy, only minor variations in the assay performance appear among the commercial immunoassay kits. However, for irregular gestations, immunoassay results can be significantly different among the kits [28]. When different forms of a protein biomarker have different molecular weights, they can be readily differentiated by mass spectrometry, resulting in more reliable measurements [29].

MALDI-TOF MS can be used alone for quantification of a protein biomarker in uncomplex biological specimens, such as urine. For example, MALDI-TOF MS has been used to semi-quantify albumin in urine for the diagnosis of albuminuria [30, 31]. This approach does not require any pretreatment of a urine sample [30], and the results are not affected by the presence of interfering substances, such as drugs, detergents, and blood, which often cause false-positive and false-negative results in conventional urinary dipstick tests [31]. Glycated and glutathionylated hemoglobin can be measured by direct MALDI-TOF MS analysis of hemolysate with both intra-assay and inter-assay CVs $<10\%$ [32]. The MALDI-TOF MS results correlated well with results obtained by using a validated routine assay for HbA1c (correlation coefficient $= 0.92$) [32].

In complex biological specimens like serum, direct MALDI-TOF MS analysis of low abundant proteins is not possible. The high and medium abundant proteins in serum will mask the signals of the targeted protein. In such a case, MALDI-TOF MS can be combined with immunoprecipitation or immunocapture techniques to enrich and unmask the signal of a protein biomarker. Because it is difficult to control the amount of the target proteins recovered from the antibody beads, stable-isotope labeled internal standard protein that has the same amino acid sequence must be added to specimens for normalizing the variations. For example, after immunoprecipitation of amyloid-beta peptides from the cerebral spinal fluid, different amyloid-beta isoforms as well as their corresponding stable-isotope labeled internal standards appear as individual peaks of expected m/z values in a MALDI-TOF mass spectrum, and their quantities can be measured with high accuracy with intra-assay CVs $<10\%$ [33]. The results obtained by this method correlated well with the results obtained by ELISA with correlation coefficients of 0.89–0.95. Using specific antibody coated beads, Mason et al. has recently developed a sensitive method for quantifying angiotensin I and angiotensin II in human plasma [34]. This assay has a limit of detection of 13 and 11 pg/mL for angiotensin I and angiotensin II, respectively. The intra-assay CVs are $<10\%$.

The limitations of MALDI-TOF MS-based quantitative analysis of large intact proteins are low specificity, low sensitivity, and low resolution. An amount of 100 fmol to 1 pmol is required for generating reliable MS signal from an intact protein. For example, a concentration of 100 fmol/μL (i.e., ~6.5 μg/mL) is required for reliable measurement of intact BSA. In addition, MALDI-TOF MS does not have good resolution to resolve large intact proteins. The accuracy of a measurement can easily be affected by the presence of protein contaminants with close molecular weights. Furthermore, wide-type proteins and corresponding mutant proteins cannot be efficiently resolved. To overcome these limitations one could digest a protein mixture first, capture the specific peptides that are commonly obtained by protease digestion (i.e., proteotypic peptides) with specific anti-peptide antibody coated beads, and finally quantify the peptides to reflect the protein concentrations. This approach is called iMALDI [35] or SISCAPA [36]. For example, epidermal growth factor receptor (EGFR) has a molecular weight of 180 kDa. The detection sensitivity of this approach for EGFR was shown to be 5 fmol [35]. If one has a MALDI-TOF/TOF MS instrument, the identity of a detected target peptide can be further confirmed by tandem MS. This could help to avoid false positive test results [35]. By using synthetic proteotypic peptides of six proteins and corresponding stable isotope peptides as internal standards for proof-of-concept, this approach has been shown to have average intra-assay CVs of 2.5% at a loading amount of 11 fmol on the sample spots [36]. Although the sensitivities of these methods are still at a magnitude of nanograms per milliliter, it is expected to be improved with the advancement of MALDI-TOF MS in the near future. Furthermore, this approach has a great potential in specific quantification of mutant proteins resulting from sense mutation of a gene sequence, e.g., EGFR with T790M mutation, which is a therapy response predictor for non-small cell lung cancer patients treated with EGFR tyrosine kinase inhibitors [37]. The major shortcoming of the SISCAPA or iMALDI approach is that it cannot differentiate different forms of a target protein if the proteotypic peptide selected for quantification does not cover the differences. For example, a proteotypic peptide lying in the N-terminal region of a target protein cannot differentiate its intact form from the C-terminal truncated forms. More details about iMALDI can be found in Chap. 6 ("Mass Spectrometry in High-throughput Clinical Biomarker Assays: Multiple Reaction Monitoring" written by Parker et al.).

2.8 *Identification of Disease Associated Aberrant Glycosylation*

There has been a long history in applying glycoprotein biomarkers for disease diagnosis and prognosis. Alternations in glycosylation changes have been observed in various diseases, such as congenital disorders of glycosylation syndrome (CDGs) [38], liver diseases [39], kidney diseases [40], and cancers [41]. A typical example of glycoprotein biomarkers for monitoring disease-associated glycosylation is circulating transferrin, which is still used in most hospitals for liver damage caused by chronic alcohol abuse [39] and identification of various types of CDGs nowadays [42]. As early as 1978, abnormal microheterogeneity of serum transferrin was observed in male alcoholics after alcohol intoxication [43]. In 1993, serum transferrin was first used to examine abnormal glycosylation in CDG patients [38]. Alternation in glycosylation of glycoproteins and glycolipids is a common feature in various cancers, and is involved in numerous ways in carcinogenesis, such as progression, cell–cell interaction, and metastasis. Tumor cells have different glycosylation machineries. Changes of glycosylation machinery in the cancer cells can be reflected in blood circulation by tracing the changes in the glycosylation of the proteins released by the tumor [44]. The poor specificity of a tumor biomarker is often due to the fact that it is also produced by normal cells under other pathological conditions. However, this problem can be reduced by measuring the circulating levels of its variants carrying cancer-associated glycosylations. For cancer diagnosis, a typical example is alpha-fetoprotein (AFP). Compared to the total serum AFP level, both fucosylated AFP and monosialylated AFP are more specific in the diagnosis of hepatocellular carcinoma (HCC) [45, 46]. Elevated mRNA expression of alpha1-6 fucosyltransferase in human HCC tissues was associated with the production of tumor-specific fucosylated AFP glycoform [47]. Serum levels of monosialylated AFP were negatively correlated with the tissue levels of beta-galactoside alpha-2,6-sialyltransferase [41].

MALDI-TOF MS can be used to identify and quantify disease-associated glycosylations carried by either a single protein or a mixture of proteins. For both cases, N-linked glycans or O-linked glycans can be cleaved from a protein preparation, cleaned up to remove interfering substances, and subsequently subjected to MALDI-TOF MS to obtain a mass spectrum of glycans (Fig. 2a). After normalization, the peak intensities of individual glycans can be used to estimate their relative levels in the preparation [5]. The intra- and interassay CVs of normalized peak intensities of *N*-glycans were reported to be $<10\%$ and $<17\%$, respectively [5, 48]. The first application of MALDI-TOF-MS to analysis of N-linked glycans on transferrin preparations (Fig. 2b) that were affinity isolated from serum samples for diagnosis of Type-I CDGs was reported in 1994 [49]. Besides analyzing glycans cleaved from glycoproteins, one could use MALDI-TOF MS to examine disease-associated glycopeptides which are obtained by proteolytic digestion of affinity isolated proteins. MALDI-TOF MS analysis of glycopeptides from serum transferrin has been applied in CDG screening system to the diagnosis of Type-II CDGs in Japan [50]. Because MALDI-TOF is a sensitive technique, only 20 μL of serum is required for screening of Type-I and Type-II CDGs [50].

When analyzing glycans released from all proteins in a tissue instead of a single protein, the concept of glycome appears. MALDI-FTICR MS was first used to obtain a semi-quantitative profile of O-linked glycome in serum, and identified potential glycan biomarkers for ovarian cancer [51]. One year later the same approach was used to discover potential glycan biomarkers for breast cancer [52]. Despite these encouraging results, a MALDI-FTICR MS instrument is far too expensive to be acquired by most of the clinical laboratories for providing routine services. Almost at the same time, another team and our team reported the use of MALDI-TOF MS for obtaining semi-quantitative profiles of serum N-linked glycome (Fig. 2c), and showed the potential use of serum N-linked glycome fingerprints in the diagnoses of metastatic prostatic cancer and liver fibrosis [5, 53]. Similar MALDI-TOF MS approaches have been used to identify serum *N*-glycan biomarkers for diagnoses of various cancers, including HCC [54], breast cancer [55], esophageal adenocarcinoma [56], and ovarian cancer [57]. In the case of breast cancer, a serum *N*-glycan at *m/z* 2,534 was found to be a potential predictor of patients' response to trastuzumab [58].

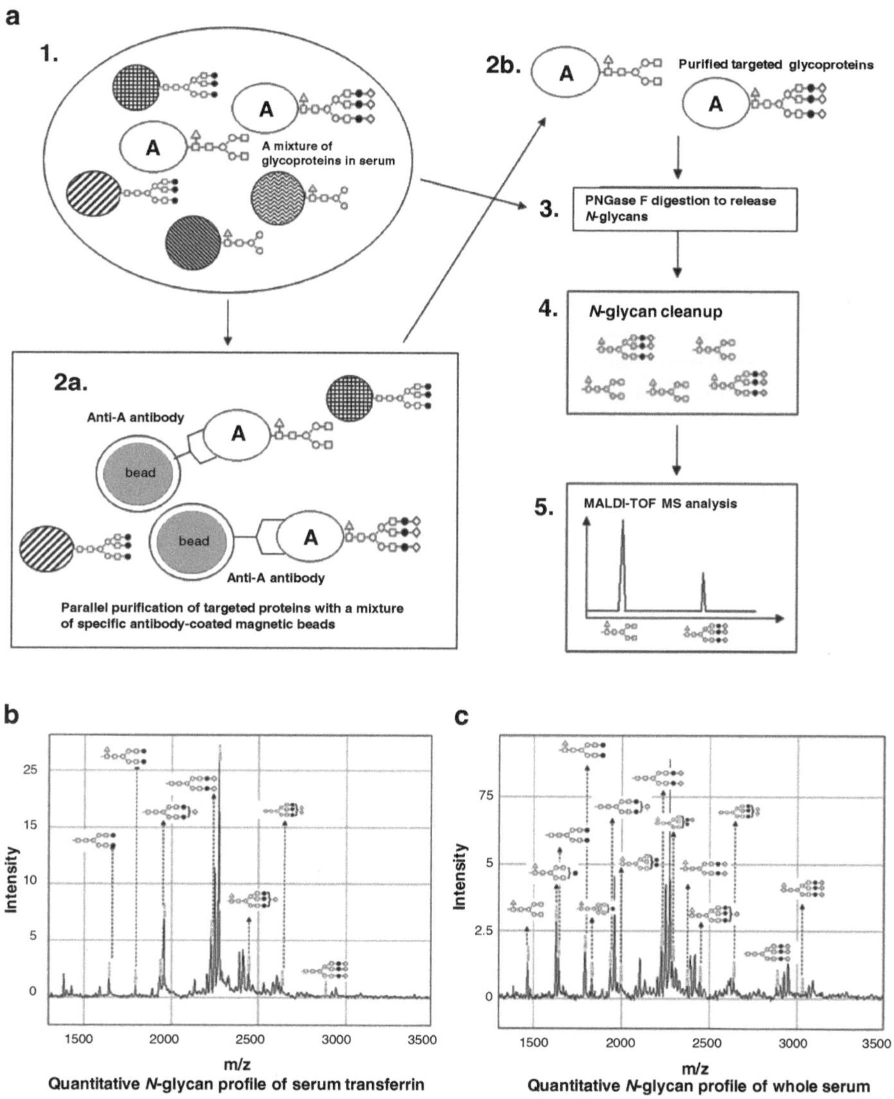

Fig. 2 (**a**) A typical workflow of quantitative profiling of N-linked glycans carried by proteins in whole serum (steps 1 and 3–5) or N-linked glycans carried by a single serum protein (steps 1–5) by MALDI-TOF MS. (**b**) Representative quantitative *N*-glycan profile from transferrin purified from serum by micro-scale antibody affinity chromatography. (**c**) Representative quantitative *N*-glycan profile from proteins in whole serum

2.9 Qualitative and Quantitative Analysis of Genetic Markers

In 1992, Nordhoff et al. first demonstrated the use of MALDI-TOF MS to detect and measure the masses of nucleic acids [59]. Three year later, the research team led by Charles Cantor showed that MALDI-TOF MS was a useful tool for DNA sequencing [60]. Since then, various MALDI-TOF MS-based methods have been being developed for applications of molecular genetics in clinical diagnosis. The successful application of MALDI-TOF MS in genotyping has been widely applied in the past 10 years. The most commonly used MALDI-TOF MS method is detection and quantification of single base primer extension products for qualitative and quantitative analysis of DNA copies containing single nucleotide polymorphism (SNP) by MALDI-TOF MS [61]. When the primers are well designed to achieve a good separation of the primer and the extension products in a mass spectrum, the genotyping assays can be combined to perform up to 15-fold multiplex SNP analysis [62, 63]. The single base primer extension assay can be applied to diagnosis and screening of hereditary diseases such as cystic fibrosis and beta-thalassemia [64, 65]. In fact, any diseases/pathological conditions that are associated with mutations in a specific gene or a specific set of genes can be easily identified and quantified by this method. This has recently been applied to the detection and quantification of the frequency of EGFR activating mutations in non-small-cell lung cancer tissues for prediction of patient's response to EGFR tyrosine kinase inhibitor [37]. This method was shown to have detection limits of 0.4–2.2% [37]. In addition, when a known amount of an oligonucleotide having a well-designed sequence is spiked into a biological sample for competition in the primer extension reaction, the primer extension method can be used for measurement of the exact number of copies of DNA containing a mutation of interest. A typical application example is detection of 60 hepatitis B virus variants in four multiplex reactions [63]. The limit of quantification was 1,000 HBV copies/mL. Besides DNA, the primer extension method can be applied to the qualitative and quantitative analysis of RNA [66]. In 2007 it was first shown that quantification of plasma placental RNA allelic ratio permitted noninvasive detection of prenatal chromosomal aneuploidy detection [67]. This work has opened a new avenue for prenatal diagnosis.

2.10 Quantification of Metabolites by MALDI-TOF MS and SALDI-TOF MS

The matrix chemical noises in the low mass region ($<m/z$ 800) make MALDI-TOF MS inferior for small molecule analysis. Despite that, attempts have been made to use MALDI-TOF MS for direct quantification of biomolecules, such as amino acids [15] and lipids [68], without the need for chemical derivatization. MALDI-TOF MS peak intensities usually increase with the amount of biomolecules. By spiking an internal standard and using an external calibration curve, one can use MALDI-TOF MS to estimate the concentration of a metabolite in a biological specimen through calculating the peak intensity ratio of the target metabolite to the internal standard. In Gogichaeva et al.'s study, methyltyrosine was used as a universal internal standard for quantification of various amino acids [15]. The calibration curves exhibited linearity in a range between 20 and 300 μM with correlation coefficients

>0.983. The between-day CVs for the majority of amino acids were <10%, with proline and arginine being exceptions with CVs of about 12% [15]. By using 4-cholesten-3-one as a universal internal standard, it was practically feasible to use MALDI-TOF MS to identify and measure the lipid composition (*m*/*z* 369.6–833.0) of VLDL, LDL and HDL [68].

It has recently been shown that by MS acquisition at the negative ion mode and using 9-aminoacridine instead of the typical chemical matrices, matrix chemical noise can be greatly reduced [69]. This method allowed the detection and quantification of metabolites having acid protons, such as amines, alcohols, carboxylic acids, phenols, and sulfonates, with high sensitivity [70]. High linearity of the MS peak intensities of the deprotonated metabolites was observed at low concentration [69, 71]. The detection limits were in the femtomole range [69, 71]. By using 9-aminoacridine as the matrix and *N*-1-napthylphthalamic acid as the universal internal standard, SPE-enriched various bile acid species from plasma specimens could be directly measured with the limit of detection within the range 0.25–4.60 μg/mL [72].

Another method for reducing matrix chemical noises is the replace of the chemical matrix by a solid matrix. This approach is called surface assisted laser desorption/ionization (SALDI) (Fig. 1) [73]. The concept of SALDI was introduced by Sunner et al. in 1995. By using graphite to replace the chemical matrix, it was shown that peptides and proteins could be detected at high sensitivities [74]. Moreover, the background signal at the low mass region was low [73]. Since then, many other solid materials, such as silicon [75], carbon nanotube [76, 77], graphene flake [78], reduced graphene oxide [79], polymer matrix [80], and gold nanoparticles [81], have been shown to be useful matrices for SALDI-TOF MS analysis of small biomolecules, including carbohydrates [76, 81], amino acids [77], and lipids [82]. On one hand, the use of a solid-phase matrix alleviates the matrix chemical noises and interference problem at the low mass range. On the other hand, it solves the problem of uneven distribution of the analytes on a sample spot. By using graphene-based materials, the MALDI-TOF mass spectra of small molecules were found to be highly reproducible [78]. The within-spot spectrum-to-spectrum CV of peak intensities for spermin was 14% for the graphene matrix, compared to 40% for the CHCA matrix [78]. Recently, Lu et al. examined the shot-to-shot and spot-to-spot reproducibility of SALDI-TOF mass spectra for polypropylene glycol polymers. The shot-to-shot and spot-to-spot CVs of the signal intensities were 1.9–7.1% and <10%, respectively [83]. SALDI-TOF MS has great potential in quantitative profiling of small biomolecules, especially metabolites.

2.11 Discovery of Metabolite Biomarkers by Quantitative Profiling

In the Post Genome Era, besides proteomics, metabolomics has been a hot topic in the past 10 years. Many research groups have been attempting to use MS technologies to obtain quantitative profiles of metabolites in patients' specimens, and to identify potential metabolite biomarkers by comparing the profiles between subjects with and without the diseases. LC-ESI MS has been the most commonly used technology in this research area [84]. ESI is a kind of atmospheric pressure ionization-based method, resulting in occurrence of ionization suppression [84]. Another disadvantage is that the use of LC limits the throughput. It can be very time consuming when one wants to obtain comprehensive metabolite profiles from over 100 biological specimens in a biomarker discovery study. Because of the high-throughput nature of MALDI-TOF MS, the use of MALDI-TOF MS in metabolite profiling is a very attractive alternative. It has been shown that 9-aminoacridine can be used to obtain quantitative cellular metabolite profiles by direct mixing of cells and the matrix without any preprocessing [69, 85]. For example, by a single direct on-spot analysis of 2,500 human acute lymphoblastic leukemia Jurkat cells, this method detected up to 150 metabolite peaks in the range of m/z 250–850 within 90 s [69]. It is important to note that signal suppression of a metabolite was observed when another metabolite with a similar chemical structure was present [86]. Hence, when using MALDI-TOF MS for quantitative analysis of metabolites, the data should be interpreted carefully. In the near future it will be interesting to see whether metabolites in plasma/serum can be directly profiled with the use of 9-aminoacridine as the matrix.

2.12 *Quantification of Metabolites by MALDI-TOF/TOF MS*

Nowadays selected reaction monitoring/multiple reaction monitoring (SRM/MRM) is the most widely accepted MS method for reliable quantification of small molecules, and typically implemented in an ESI triple quadrupole (ESI-QqQ) mass spectrometer (see Chap. 6 for details of the basic principle and instrumentation). The QqQ tandem mass analyzer is used dedicatedly in the SRM/MRM method because a quadrupole mass analyzer can be used as a mass filter, which only allows charged molecules of a specific m/z value to pass through the mass analyzer for either subsequent fragmentation or detection. By undertaking the filtering approach, the background noise can be greatly reduced, leading to high detection sensitivity. SRM cannot be implemented in MALDI-TOF/TOF MS instruments. Few reports on using MALDI-TOF/TOF MS in tandem MS mode for quantification have been available. Gogichaeva et al. showed that amino acids could be fragmented by MALDI-TOF/TOF MS [87]. By calculating the peak intensity ratios of the indicator fragment ions of the target amino acids to the indicator fragment ion of an internal standard, good correlation between the mixture component molar ratios and indicator fragment ions intensity ratios was observed [87]. Although correlation coefficients and coefficients of variation of their MALDI-TOF/TOF MS method were not reported, the study highlighted the potentials of applying MALDI-TOF/TOF MS to biomolecule quantification [87].

Recently, using citrulline for proof-of-concept, our team has developed a novel MALD-TOF/TOF MS-based quantification method called parallel fragmentation monitoring (PFM) [88]. This method is comparable to SRM. As in the SRM method, the PFM method also requires at least two pairs of precursor and selected fragment ions of specific *m*/*z* values, one pair for the target molecule and one pair for the internal standard. A stable isotope analog of the target molecule only 1 mass unit heavier is used as an internal standard, so that precursor ions of both the target molecule and internal standard can be specifically isolated with the first TOF analyzer at the same time, and undergo fragmentation simultaneously to yield a full range composite MS/MS spectrum. In both the SRM and PFM methods, the peak area ratio of the selected fragments of the target analyte to internal standard was used for quantification. The use of a stable isotope analog should also be able to minimize the error due to systematic bias of the instrumentation, and normalize the recovery yield after enriching the analytes from the biological samples for quantification. To reduce the matrix noises in the low mass range, a carbon-based nanomaterial was used as the matrix. The performance of the PFM method appears to be comparable to those of the SRM/MRM methods. Both PFM and SRM/MRM methods generated linear calibration curves with correlation coefficients >0.99 (Fig. 3) [88–90]. Moreover, both types of assays gave the within- and between-day CVs $\leq 10\%$ [88–90]. Our results also showed that the calibration curves were highly reproducible. Daily calibration or use of a stored calibration generated highly similar measurement values [88]. This suggests that PFM can potentially be a cost and time effective and robust technology for quantification of biomolecules in routine clinical chemistry laboratories.

The major advantage of using MALDI-TOF/TOF MS instead of MALDI-TOF MS for quantification is that MALDI-TOF/TOF MS has higher detection specificity and sensitivity for direct quantification of a target biomolecule in a complex biological specimen or in a partially enriched preparation. MALDI-TOF MS does not have enough resolution to resolve two ion species with highly close molecular weights, but they can be easily differentiated by looking at the fragmentation pattern. Even a highly advanced MALDI-TOF MS with ultra-high resolution, such as MALDI-FTICR MS, is not able to differentiate the naturally occurring isomers, such as leucine and isoleucine, by only focusing on the intact ions because of the exactly similar molecular weights. Isomers can only be differentiated on the basis of fragment ions. Furthermore, the background noises and interference from the other biomolecules in the preparation, like signal suppression by biomolecules sharing similar chemical structures, can be minimized by measuring ratios of the indicator fragment ions, resulting in higher detection sensitivity and measurement accuracy.

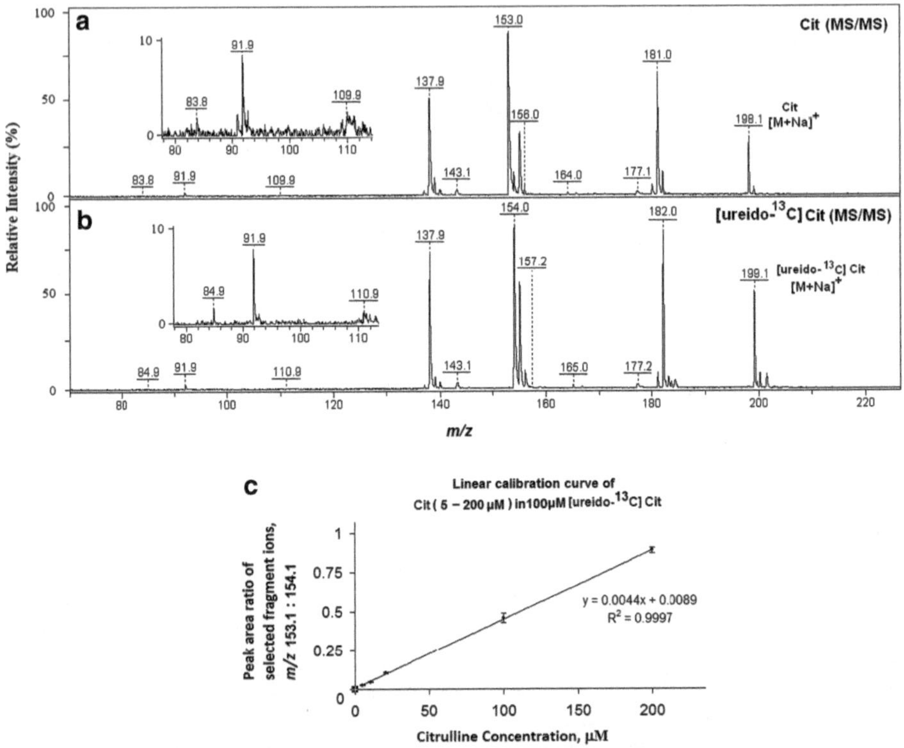

Fig. 3 Representative MALDI-TOF/TOF tandem MS spectra of citrulline (**a**) and [ureido-^{13}C] citrulline (**b**) acquired independently with graphene flake as the solid matrix. (**c**) Representative linear calibration curve of the PFM assay for quantitative analysis of citrulline in the range of 10–250 μM. The peak intensity ratios of the indicator fragment ion of citrulline (*m/z* 153.1) to that of [ureido-^{13}C] citrulline (*m/z* 154.1) in the calibration standards were plotted against the citrulline concentrations

3 MALDI-QqQ MS

3.1 Quantification of Biomarkers

Although a QqQ tandem mass analyzer is usually coupled with an ESI source, it can also be coupled with a MALDI source [91]. In MALDI-QqQ MS the identity of a biomolecule can be defined by a mass transition ion pair as in the case of SRM [91]. By operating the QqQ analyzer as mass filters for the targeted precursor ions and fragment ions, the matrix chemical noises at the low mass region can be greatly reduced [91]. Comparing the quantitative results obtained by MALDI-QqQ MS and ESI-QqQ MS for 53 small-molecule pharmaceutical compounds, Gobey et al. demonstrated the potentials of MALDI-QqQ MS for high-throughput quantification of small biomolecules [91]. When operating MALDI-QqQ MS in SRM/MRM mode, the CVs for quantifications of small biomolecule or drug are typically around 10% [91–94]. It has recently been shown that MALDI-QqQ MS can also be used to measure protein biomarkers in plasma by quantifying their proteotypic peptides in the presence of corresponding isotopically labeled peptide standards, as in case of typical MRM methods [95]. The measurement results are highly comparable to those obtained by using ESI-QqQ MS. Both technologies are accurate (within-day CVs <20%) and precise (relative errors <20%) for protein quantification [95]. Because MALD-QqQ MS is a relatively new technology, the currently available data have been limited. However, all the recent reports have demonstrated that MALDI-QqQ MS in SRM/MRM mode, which combines the merits of MALDI ionization technology and those of the conventional SRM/MRM approach, is a reliable high-throughput technology for biomolecule quantification. More successful applications of MALDI-QqQ MS to quantification of biomarkers in human specimens should be forthcoming.

4 SELDI-TOF MS

4.1 Basic Principle

Surface-enhanced laser desorption/ionization TOF mass spectrometry (SELDI-TOF MS) is a variant of MALDI-TOF MS, and is mainly designed for quantitative analysis of proteins in biological samples. This concept was first introduced by Hutchen and Yip in 1993 [96]. Instead of spotting of a mixture of proteins on a MALDI sample plate, a mixture of proteins is subjected to ProteinChip array-based retentate chromatography before MALDI-TOF analysis. ProteinChip arrays coated with different types of chromatographic materials (hydrophilic, hydrophobic, cationic exchange, anionic exchange, immobilized metal affinity, antibody affinity, ligand affinity, etc.) can selectively bind and concentrate proteins with the matched physicochemical or biochemical properties (Fig. 4a, b). Those nonspecifically bound proteins and impurities are then washed away with suitable washing buffers [97]. Retained proteins are finally co-crystallized with a chemical matrix (Fig. 4c), and subjected to MALDI-TOF MS for unbiased detection of protonated proteins (Fig. 4d, e). In the case of SELDI-TOF MS, sinapinic acid is the most commonly used matrix to assist desorption/ionization of proteins. Most of the proteins are detected as singly charged protonated molecules, and presented as individual peaks in a mass spectrum, resulting in a proteomic profile (Fig. 4f). The combinations of specific *m/z* values and the physicochemical properties that are reflected by the type of the ProteinChip arrays used provide unique identities for individual proteins [97]. This is why SELDI-TOF MS is commonly regarded as a proteomic fingerprinting technology. A series of follow-up experimental work is needed to purify the corresponding proteins and decipher the true identities [98–100].

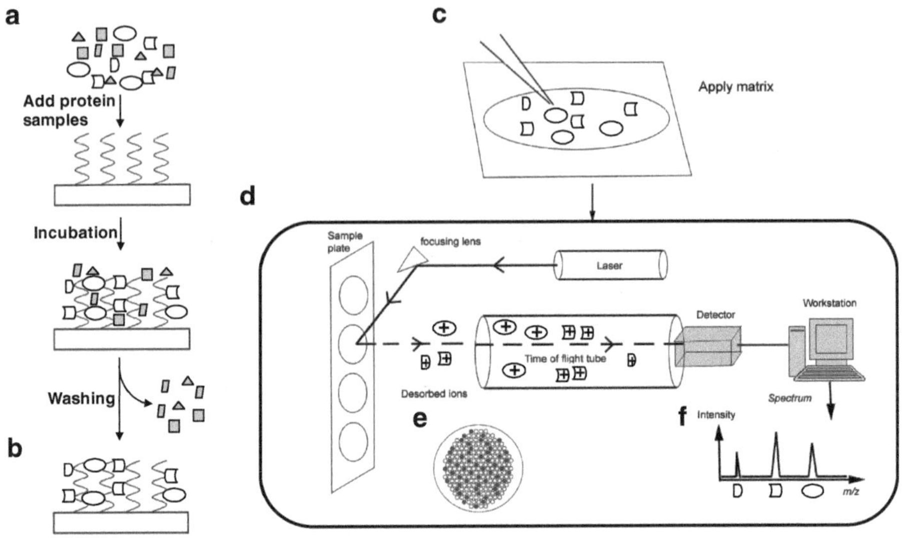

Fig. 4 (**a**) Typical workflow of quantitative proteomic profiling by SELDI-TOF MS. After denaturation and dilution, a patient sample is added to a binding surface of a ProteinChip array. (**b**) After incubation and washing, proteins with matched physicochemical/biochemical properties are retained on the surface. (**c**) Then a chemical matrix is added. (**d**) After drying, the sample spot is subjected to MALDI-TOF MS analysis to obtain signals from various positions with a regular spacing (**e**) (*dark circles*) throughout the entire spot. (**f**) After summing up the MS signals, a quantitative proteomic profile is obtained

It is worth noting that the concepts of SALDI and SELDI can be combined. A solid-phase matrix can both capture biomolecules with a particular physicochemical property and assist desorption/ionization of the captured biomolecules in MALDI-TOF MS analysis. Immobilization of CHCA onto the hydrophobic ProteinChip arrays allowed direct quantitative profiling of urine proteins by SELDI-TOF MS without the need for adding any chemical matrix after sample binding and washing [101]. Recently, a graphene-based SELDI probe has been developed for capture and direct detection of DNA oligomer without addition of any chemical matrix [102].

4.2 Quantitative Issues in the SELDI-TOF Mass Spectra

As in the case of MALDI-TOF MS, with appropriate MS analysis conditions, SELDI-TOF MS is quantitative. On the ProteinChip arrays, chromatographic resins are coated on a film of hydrogel, which provides a network for the formation of fine analyte-matrix co-crystals. Furthermore, in a typical SELDI-TOF MS experiment, an unbiased automated MS acquisition strategy is used. MS signals from 60 to 120 laser shots are obtained from a sample spot in a linear sweep or from various positions with a regular spacing on the entire sample spot, and are summed up to form a representative mass spectrum. After normalizing the MS signals by either total ion current and/or total peak intensity, the intensity values of the protein/peptide peaks is highly reproducible. The intra-assay and inter-assay CVs for the normalized intensities of majority of the SELDI peaks are between 5% and 25% [103, 104]. With the use of standardized experimental protocol and quality control strategy, the inter-laboratory CVs of the normalized peak intensities are between 15% and 36% [105]. By using a combination of ProteinChip arrays with different chromatographic coatings, SELDI-TOF MS can be used to obtain comprehensive semi-quantitative profiles of proteins with molecular weights between 2 and 250 kDa [106].

4.3 General Biomedical Applications of SELDI-TOF MS

Similar to other affinity technologies, SELDI-TOF MS can be applied to various types of research projects where appropriate. It all depends on what chromatographic functional groups, affinity materials, or proteins are being conjugated covalently on the ProteinChip array surface. It can be used to capture and profile transcription factors by coating with DNA materials of a specific sequence [107]. It can also be used to study the effect of DNA methylation on binding the transcriptional factors to a DNA sequence [108]. Such an approach could help to characterize transcription factors and to screen for differences in cellular regulatory networks. When a specific protein is coated, it could be used to study protein–protein interaction [109, 110]. For example, it has been used to search for protein–protein interaction partners for S100A8 [109] and GlialCAM [110]. When the ProteinChip arrays are coated with a specific antibody, specific protein or protein complex can be purified for subsequent analysis. After a specific protein has been captured, SELDI-TOF MS can identify and provide quantitative information of individual variants. They could be structural variants with different amino acid compositions, e.g., amyloid beta peptide variants [111] and SAA variants [112], as well as variants with different post-translational modifications, e.g., S-glutathionylated and S-cysteinylated variants of transthyretin [113] and glycosylated variants of eosinophil cationic protein [114]. Above all, SELDI-TOF MS has been more commonly used for quantitative profiling of biological samples to search for protein biomarkers. Theoretically, SELDI-TOF MS can also be applicable to high-throughput quantification of other types of biomolecules, like metabolites and glycans. In the following sections the applications of SELDI-TOF MS to protein biomarker discovery will be reviewed in more detail.

4.4 High-Throughput Technology for Biomarker Discovery

In a typical SELDI-TOF MS experiment, biological specimens can be directly added on the ProteinChip arrays without any preprocessing, or only after several simple steps for denaturation and dilution. The ProteinChip arrays can be assembled in a 96-well plate format, and the binding and washing procedures can be performed as if carrying out an enzyme-linked immunosorbent assay. The use of the ProteinChip arrays has greatly simplified the protein profiling assay workflow. Another advantage of SELDI-TOF MS is its capacity for high throughput during mass spectrum acquisition, as in the case of MALDI-TOF MS. In addition, the combinations of specific *m/z* values and the type of ProteinChip arrays used provide unique identities for individual proteins. These are the major reasons why SELDI-TOF MS has been widely used for analysis of biological specimens, especially for biomarker discovery, since it first appeared as a commercially available platform in 1997. As of today, there are at least 850 publications on applications of SELDI-TOF MS to protein biomarker discovery. It has been used to analyze a wide range of biological samples, for example, serum [100, 103–106], plasma [110, 113, 115], urine [101, 113, 116], tears [117–119], cerebrospinal fluid [120–122], amniotic fluid [123–125], tissue/cell lysate [107, 126–128], etc. It has been applied to the discovery of potential biomarkers for various types of diseases, for example, cancers [103–106, 112, 129], infectious diseases [99, 100, 115, 130], autoimmune diseases [118, 131], eye diseases [117, 119], neurological diseases [120–122], perinatal and neonatal diseases [123–126], etc. In a typical proteomic profiling experiment the individual biological samples were first denatured with urea and non-ionic solvent to denature or destroy the non-covalent protein–protein interaction, and then diluted in an appropriate binding buffer for subsequent SELDI-TOF MS analysis [99, 100, 103–106, 112, 115, 129]. In such an approach only a very small amount of biological sample is needed. For serum/plasma specimens, as little as 2 μL of serum samples would be enough [99, 129, 130].

One reason for the popularity of SELDI-TOF MS in biomarker discovery is that it is complementary to two-dimensional polyacrylamide gel electrophoresis (2D-PAGE) for the quantitative analysis of intact proteins or their fragments formed in vivo. The most important point is that 2D-PAGE is the best for resolving proteins with molecular weight in the range of 10–250 kDa, while SELDI-TOF MS has the best resolving range of 1–20 kDa. Furthermore, highly hydrophobic proteins, such as membrane proteins, and proteins with isoelectric points (p*I*) less than 3 and higher than 11, are usually poorly resolved by 2D-PAGE, but can be satisfactorily analyzed by SELDI-TOF MS.

Another reason for the popularity of SELDI-TOF MS in biomarker discovery is that SELDI-TOF MS itself has a great potential to be used as a clinical diagnostic tool. The simplicity in the assay procedure and its short turn-around time allow this to be implemented in routine clinical chemistry laboratories. We could easily translate the laboratory research findings into clinical use. Diagnosis/prognosis could be based on the intensity of a SELDI peak at a specific *m/z* value and the usage of a specific type of ProteinChip arrays, or based on a combination of specific protein peaks which have been identified with the use of several specific types of ProteinChip arrays.

4.5 Host Response Proteins Forming the SELDI Proteomic Fingerprints

Proteins corresponding to the diagnostic/prognostic peaks could be purified by micro-scale chromatography with the same binding condition, separated by gel electrophoresis, and finally identified by using typical approaches, e.g., peptide mass fingerprint, tandem MS, etc. With clear protein identities, specific immunoassays could be developed. Now it is clear that the majority of disease-associated proteomic fingerprints are composed of intact forms, fragments, and/or post-translationally modified forms of host response proteins, such as apolipoprotein A1, apolipoprotein A2, apolipoprotein C1, apolipoprotein C2, apolipoprotein C3, alpha-1 antichymotrypsin, complement component 3a, complement component 3c, fibrinogen, immunoglobulin kappa light chain, inter-alpha trypsin inhibitor heavy chain 4, haptoglobin, beta-2 microglobulin, platelet factor 4, SAA, transthyretin, beta-thromboglobulin, etc. [99, 100, 132–139]. In fact, host response proteins were also identified as potential biomarkers when the serum/plasma proteomic profiles were compared by using other techniques, such as 2D-PAGE [140, 141], magnetic beads-based MALDI-TOF MS [24, 25], and even shotgun proteomic profiling by LC-ESI-MS [142–144]. The use of signatures of host-response proteins as disease biomarkers has both pros and cons. The major advantage is that the host response of a patient helps to amplify the signal for the presence of a particular disease, which may help to identify a disease at an early stage [132]. The major disadvantage is that the specificity of those host-response signatures should be carefully validated before they can be claimed as disease-specific biomarkers. Similar symptoms, which generate specific host-response protein signatures, can easily be found in other diseases.

4.6 Presence of Systemic Bias in Biomarker Discovery Studies

Although SELDI-TOF MS has been a popular technology for biomarker discovery, there have been doubts about the reliability of this technology. Before discussing this issue, I would like to emphasize this should not be a problem that is only restricted to the SELDI proteomic profiling studies. Such a problem has been observed in many SELDI-TOF MS studies. It may be because SELDI-TOF MS was the first high-throughput technology that allowed quantitative profiling and comparison of the serum/plasma proteins in a large number of patient samples within a very short period of time. After much more serum proteomic/metabolomic profiling, data obtained by using other technologies are available, I believe that similar problems will be observed. In this section, selected biomarker discovery studies employing SELDI-TOF MS technology will be used as examples for reviewing this issue.

The typical example is the application of SELDI-TOF MS to the diagnosis of ovarian cancer. In 2002, Petricoin et al. identified a pattern of SELDI peaks that could completely differentiate ovarian cancer cases from non-cancer cases in the training set. For the masked set, the diagnostic pattern achieved a sensitivity of 100% and specificity of 95% [145]. Unfortunately, when the data set was reanalyzed by other teams, it was found that there was significant non-biological experimental bias between the cancer and control subjects [146], and the features in the noise regions of the SELDI mass spectra allowed discrimination of control subjects from cancer patients [147, 148]. These analysis results suggested that (1) the cancer and control samples had been analyzed separately and (2) there was a change in the experimental protocol in the middle of the study.

Another important example is the identification of SELDI peaks for diagnosis of prostate cancer. In 2002, by using the copper(II) ion loaded metal affinity (Cu^{2+}-IMAC) ProteinChip Array, a pilot single-center study showed that serum protein fingerprinting was useful for diagnosis of prostate cancer [149]. When classifying the blind test samples, the sensitivity and specificity were found to be 83% and 97%, respectively. Subsequently, a series of follow-up studies were performed to validate the value of serum proteomic profiling with Cu^{2+}-IMAC ProteinChip arrays in the diagnosis of prostate cancer. To allow validation carried out by six research centers, a standard protocol and quality control system was developed [105]. Then serum samples (181 prostate cancer patients, 143 benign prostatic hyperplasia cases, and 220 normal controls, who were age and race-matched) from Eastern Virginia Medical School (EVMS) were used to construct a decision algorithm for classifying 42 prostate cancer patients and 42 normal controls provided by four institutions [150]. All test samples were distributed to six laboratories for analysis. The final conclusion was that the decision algorithm was unsuccessful in separating cancer from controls. Analysis of the experimental data for biomarker discovery indicated that the sample source is the major factor affecting the results.

4.7 *Overcoming Systemic Bias in Biomarker Discovery Studies*

Inappropriate selection of control subjects (i.e., selection bias) is one of the major causes of systemic bias. Selection bias and information bias will appear when the diseased and control subjects were recruited from two different populations, such as two different clinics. For example, three laboratories had attempted to use SELDI-TOF MS to identify the biomarkers for detection of severe acute respiratory syndrome (SARS) in adults [99, 151, 152]. In two of the three studies, controls cases were recruited from other clinics [151, 152]. Patients with other types of respiratory infections had been included as the controls. SAA concentration was found to be significantly higher in the SARS patients than in the controls. One study included SAA into the diagnostic model for detection of SARS [152]. In the third study which was performed by our team, both SARS patients and control subjects were recruited from the same clinics. The control subjects were suspected SARS cases, but were later proven to be negative for SARS [99]. In this study, both the SELDI-TOF MS assay and immunoassay showed that SAA was elevated in the SARS patients [153]. However, SAA levels were found to be much higher in the control group, indicating that SAA was not a useful biomarker for diagnosis of SARS [153]. These three studies clearly illustrate the importance of recruiting the diseased and control cases from the same clinic.

For case–control biomarker study, confounding bias should also be controlled. If it is not controlled, the biomarkers found could be related to the characteristics of the disease group, but not related to the disease itself. Patients with gastroenterological cancer may lose appetite, leading to under nutrition [154]. Malnutrition will become one of confounding factors, and some differential SELDI peaks could be related to the nutritional status. For hepatitis virus-related liver cancer, it is well known that gender is one of the confounding factors [155]. Smoking is a well-known confounding factor for lung cancer [156]. Levels of considerable amounts of blood proteins can be changed in response to smoking [157]. For biomarker discovery, even though the diseased and control cases are recruited from the same clinics, unknown confounding factors still exist. In our recent gastric cancer study we attempted to use post-operative serum samples to verify the validity of the potential proteomic markers found by comparing the diseased and control cases from the same clinics [129]. Surprisingly, over 80% of the potential biomarkers could not show a reverse in the serum levels after the removal of the tumors from the patients. This strongly suggested that most of the differential SELDI peaks between the gastric cancer and control groups were not specifically associated with gastric cancer, but only associated with certain characteristics of the gastric cancer patients. In our SARS study we identified the clinical and biochemical variables which were significantly altered in the SARS patients, and attempted to verify the potential diagnostic SELDI peaks by only considering those that were significantly correlated with at least two disease-associated biochemical/clinical parameters as SARS-specific (Fig. 5a) [99]. Similar to the gastric cancer study, about 80% of the differential SELDI peaks were rejected in the SARS study. Both the gastric cancer study and the SARS study have highlighted the high risk of false discovery when we simply consider the differential SELDI peaks as potential biomarkers. The presence of about 80% of the differential SELDI peaks, which are likely caused by confounding bias, are not restricted to the studies employing SELDI-TOF MS. Our group recently attempted to identify circulating host response biomarkers for diagnosis of late-onset sepsis or necrotizing enterocolitis in preterm infants suspected for the diseases by using hydrophobic magnetic beads and MALDI-TOF MS [25]. By using the longitudinal samples to verify the clinical relevance of the differential proteomic features, again about 80% of them were rejected (Fig. 5b). Encouragingly, the diagnostic values of the verified proteomic features were subsequently confirmed in the prospective study [25].

While one can reduce the systemic bias in a single center study by verification with longitudinal samples or by correlation with known disease-associated changes, one can also reduce the systemic bias by using samples from multiple centers [158].

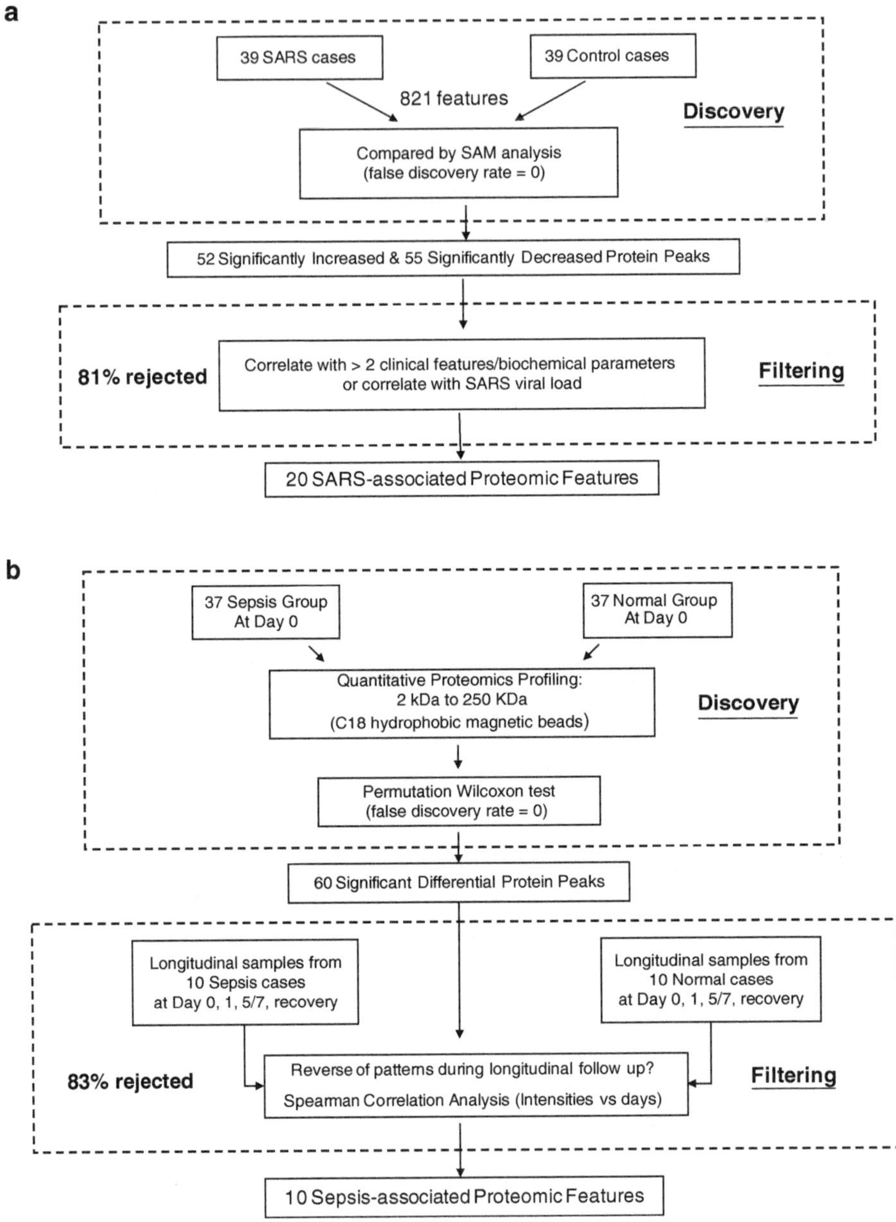

Fig. 5 The study designs which were used to identify biomarkers for diagnosis of SARS in adults (**a**) and diagnosis of necrotizing enterocolitis/late-onset sepsis in preterm infants (**b**) by undertaking MS-based proteomic profiling approaches. In the SARS study, potential diagnostic SELDI peaks were filtered by only considering those that were significantly correlated with at least two disease-associated biochemical/clinical parameters as SARS-specific (**a**) [99]. In the preterm infant study, the longitudinal samples were used to verify the clinical relevance of the differential proteomic features [25]. Only the differential MS peaks showing statistically significant reverse of peak intensities upon recovery were retained (**b**). In both studies, about 80% of differential SELDI peaks, which were obtained by case–control comparison, were rejected

Multi-center design provides an unbiased clinical validation of the proteomic diagnostic models. Zhang and Chan have proposed a multicenter design that helps to eliminate the systemic biases in samples and site-associated confounding variables in biomarker discovery [159]. In the biomarker discovery phase, cases from independent sites are used separately and independently to identify the potential biomarkers. The potential biomarkers from the different sites are cross-compared to produce a common set. In the validation phase, the clinical value of the common set is further validated using independent samples from additional sites. By using this multicenter design, a panel of ovarian cancer-associated protein biomarkers that were identified in blood samples by SELDI-TOF MS finally became the first in vitro diagnostic multivariate index assay (IVDMIA) of proteomic biomarkers, which was recently cleared by the US FDA (Food and Drug Administration) [158, 160, 161].

5 Future Prospectives

The global research efforts on the development and biomedical applications of MALDI-based technologies in the past 27 years have shown great promise in facilitating biomarker discovery and in clinical diagnostic applications. The concepts of MALDI, SALDI, SELDI, and PFM are complementary to each other. Theoretically, all these technologies can be combined, leading to the next generation of MALDI MS technologies. Although SELDI-TOF MS is commonly regarded as a proteomic fingerprinting technology, SELDI-TOF MS should also be applicable to quantitative profiling of other types of biomolecules, such as glycans and metabolites for biomarker discovery or identification of diagnostic fingerprints. Furthermore, SELDI can be coupled with TOF/TOF MS. Then targeted quantification of specific metabolites, small proteins and proteotypic peptides from large proteins, can be achieved by undertaking the PFM approach, while enrichment/purification procedures are much simplified. When the binding surface of a SELDI chip is made of materials that can also assist laser desorption and ionization process (i.e., combination of SELDI and SALDI), a SELDI-TOF/TOF MS setup will become an instrument for cost-effective measurement of biomarkers with ultrahigh throughput and high detection sensitivity and specificity. Ultimately, when a TOF or TOF/TOF analyzer can be miniaturized to a portable size without sacrificing resolution, medical diagnostic applications of MALDI-based technologies at the bedside or even at home will be become possible.

Acknowledgements The authors are grateful for the continuous support from the Li Ka Shing Foundation on developments and applications of mass spectrometry technologies to biomedical research.

References

1. Karas M, Bachmann D, Hillenkamp F (1985) Influence of the wavelength in high-irradiance ultraviolet laser desorption mass spectrometry of organic molecules. Anal Chem 57: 2935–2939
2. Tanaka K, Waki H, Ido Y, Akita S, Yoshida Y, Yoshida T (1988) Protein and polymer analyses up to m/z 100 000 by laser ionization time-of-flight mass spectrometry. Rapid Commun Mass Spectrom 2:151–153
3. Zenobi R, Knochenmuss R (1998) Ion formation in MALDI mass spectrometry. Mass Spectrom Rev 17:337–366
4. Pang RTK, Johnson PJ, Chan CML, Kong EKC, Chan ATC, Sung JJY, Poon TCW (2004) Technical evaluation of MALDI-TOF mass spectrometry for quantitative proteomic profiling – matrix formulation and application. Clin Proteomics 1:259–270
5. Kam RKT, Poon TCW, Chan HLY, Wong N, Hui AY, Sung JJY (2007) High-throughput quantitative profiling of serum N-glycome by MALDI-TOF mass spectrometry and N-glycomic fingerprint of liver fibrosis. Clin Chem 53:1254–1263
6. Roskey MT, Juhasz P, Smirnov IP, Takach EJ, Martin SA, Haff LA (1996) DNA sequencing by delayed extraction-matrix-assisted laser desorption/ionization time of flight mass spectrometry. Proc Natl Acad Sci USA 93:4724–4729
7. Guilhaus M (1995) Principles and instrumentation in time-of-flight mass spectrometry. J Mass Spectrom 30:1519–1532
8. Medzihradszky KF, Campbell JM, Baldwin MA, Falick AM, Juhasz P, Vestal ML, Burlingame AL (2000) The characteristics of peptide collision-induced dissociation using a high-performance MALDI-TOF/TOF tandem mass spectrometer. Anal Chem 72:552–558
9. Suckau D, Resemann A, Schuerenberg M, Hufnagel P, Franzen J, Holle A (2003) A novel MALDI LIFT-TOF/TOF mass spectrometer for proteomics. Anal Bioanal Chem 376: 952–965
10. Ho CS, Lam CW, Chan MH, Cheung RC, Law LK, Lit LC, Ng KF, Suen MW, Tai HL (2003) Electrospray ionisation mass spectrometry: principles and clinical applications. Clin Biochem Rev 24:3–12
11. Thiede B, Höhenwarter W, Krah A, Mattow J, Schmid M, Schmidt F, Jungblut PR (2005) Peptide mass fingerprinting. Methods 35:237–247
12. Kapp E, Schütz F (2007) Overview of tandem mass spectrometry (MS/MS) database search algorithms. Curr Protoc Protein Sci (Chapter 25:Unit25.2)
13. Krutchinsky AN, Chait BT (2002) On the nature of the chemical noise in MALDI mass spectra. J Am Soc Mass Spectrom 13:129–134
14. van Kampen JJ, Burgers PC, de Groot R, Gruters RA, Luider TM (2011) Biomedical application of MALDI mass spectrometry for small-molecule analysis. Mass Spectrom Rev 30:101–120
15. Alterman MA, Gogichayeva NV, Kornilayev BA (2004) Matrix-assisted laser desorption/ ionization time-of-flight mass spectrometry-based amino acid analysis. Anal Biochem 335: 184–191
16. Fuchs B, Süss R, Schiller J (2010) An update of MALDI-TOF mass spectrometry in lipid research. Prog Lipid Res 49:450–475
17. Goetz JA, Novotny MV, Mechref Y (2009) Enzymatic/chemical release of O-glycans allowing MS analysis at high sensitivity. Anal Chem 81:9546–9552
18. Galesio M, Nuñez C, Diniz MS, Welter R, Lodeiro C, Luis Capelo J (2012) Matrix-assisted laser desorption/ionization time of flight spectrometry for the fast screening of oxosteroids using aromatic hydrated hydrazines as versatile probe. Talanta 100:262–269
19. Cheng AJ, Chen LC, Chien KY, Chen YJ, Chang JT, Wang HM, Liao CT, Chen IH (2005) Oral cancer plasma tumor marker identified with bead-based affinity-fractionated proteomic technology. Clin Chem 51:2236–2244

20. Freed GL, Cazares LH, Fichandler CE, Fuller TW, Sawyer CA, Stack BC Jr, Schraff S, Semmes OJ, Wadsworth JT, Drake RR (2008) Differential capture of serum proteins for expression profiling and biomarker discovery in pre- and posttreatment head and neck cancer samples. Laryngoscope 118:61–68
21. Pakharukova NA, Pastushkova LK, Trifonova OP, Pyatnitsky MA, Vlasova MA, Nikitin IP, Moshkovsky SA, Nikolayev EN, Larina IM (2009) Optimization of serum proteome profiling of healthy humans. Hum Physiol 35:350–356
22. Sui W, Dai Y, Zhang Y, Chen J, Liu H, Huang H (2012) Proteomic profiling of nephrotic syndrome in serum using magnetic bead based sample fractionation & MALDI-TOF MS. Indian J Med Res 135:305–311
23. Jimenez CR, El Filali Z, Knol JC, Hoekman K, Kruyt FA, Giaccone G, Smit AB, Li KW (2007) Automated serum peptide profiling using novel magnetic C18 beads off-line coupled to MALDI-TOF-MS. Proteomics Clin Appl 1:598–604
24. Wong MYM, Yu KOY, Poon TCW, Ang IL, Law MK, Chan KYW, Ng EWY, Ngai SM, Sung JJY, Chan HLY (2010) A magnetic bead-based serum proteomic fingerprinting method for parallel analytical analysis and micropreparative purification. Electrophoresis 31: 1721–1730
25. Ng PC, Ang IL, Chiu RW, Li K, Lam HS, Wong RP, Chui KM, Cheung HM, Ng EW, Fok TF, Sung JJ, Lo YM, Poon TC (2010) Host-response biomarkers for diagnosis of late-onset septicemia and necrotizing enterocolitis in preterm infants. J Clin Invest 120:2989–3000
26. Villanueva J, Martorella AJ, Lawlor K, Philip J, Fleisher M, Robbins RJ, Tempst P (2006) Serum peptidome patterns that distinguish metastatic thyroid carcinoma from cancer-free controls are unbiased by gender and age. Mol Cell Proteomics 5:1840–1852
27. Villanueva J, Nazarian A, Lawlor K, Yi SS, Robbins RJ, Tempst P (2008) A sequence-specific exopeptidase activity test (SSEAT) for "functional" biomarker discovery. Mol Cell Proteomics 7:509–518
28. Cole LA (1997) Immunoassay of human chorionic gonadotropin, its free subunits, and metabolites. Clin Chem 43:2233–2243
29. Lund H, Torsetnes SB, Paus E, Nustad K, Reubsaet L, Halvorsen TG (2009) Exploring the complementary selectivity of immunocapture and MS detection for the differentiation between hCG isoforms in clinically relevant samples. J Proteome Res 8:5241–5252
30. Shiea J, Cho YT, Lin YH, Chang CW, Lo LH, Lee YC, Ke HL, Wu WJ, Wu DC (2008) Using matrix-assisted laser desorption/ionization time-of-flight mass spectrometry to rapidly screen for albuminuria. Rapid Commun Mass Spectrom 22:3754–3760
31. Cho YT, Chen YS, Hu JL, Shiea J, Yeh SM, Chen HC, Lee YC, Wu DC (2012) The study of interferences for diagnosing albuminuria by matrix-assisted laser desorption ionization/time-of-flight mass spectrometry. Clin Chim Acta 413:875–882
32. Biroccio A, Urbani A, Massoud R, di Ilio C, Sacchetta P, Bernardini S, Cortese C, Federici G (2005) A quantitative method for the analysis of glycated and glutathionylated hemoglobin by matrix-assisted laser desorption ionization-time of flight mass spectrometry. Anal Biochem 336:279–288
33. Gelfanova V, Higgs RE, Dean RA, Holtzman DM, Farlow MR, Siemers ER, Boodhoo A, Qian YW, He X, Jin Z, Fisher DL, Cox KL, Hale JE (2007) Quantitative analysis of amyloid-beta peptides in cerebrospinal fluid using immunoprecipitation and MALDI-Tof mass spectrometry. Brief Funct Genomic Proteomic 6:149–158
34. Mason DR, Reid JD, Camenzind AG, Holmes DT, Borchers CH (2012) Duplexed iMALDI for the detection of angiotensin I and angiotensin II. Methods 56:213–222
35. Jiang J, Parker CE, Hoadley KA, Perou CM, Boysen G, Borchers CH (2007) Development of an immuno tandem mass spectrometry (iMALDI) assay for EGFR diagnosis. Proteomics Clin Appl 1:1651–1659
36. Anderson NL, Razavi M, Pearson TW, Kruppa G, Paape R, Suckau D (2012) Precision of heavy-light peptide ratios measured by MALDI-Tof mass spectrometry. J Proteome Res 11: 1868–1878

37. Su KY, Chen HY, Li KC, Kuo ML, Yang JC, Chan WK, Ho BC, Chang GC, Shih JY, Yu SL, Yang PC (2012) Pretreatment epidermal growth factor receptor (EGFR) T790M mutation predicts shorter EGFR tyrosine kinase inhibitor response duration in patients with non-small-cell lung cancer. J Clin Oncol 30:433–440
38. Stibler H, Jaeken J (1990) Carbohydrate deficient serum transferrin in a new systemic hereditary syndrome. Arch Dis Child 65:107–111
39. Stibler H (1991) Carbohydrate-deficient transferrin in serum: a new marker of potentially harmful alcohol consumption reviewed. Clin Chem 37:2029–2037
40. Barratt J, Smith AC, Feehally J (2007) The pathogenic role of IgA1 O-linked glycosylation in the pathogenesis of IgA nephropathy. Nephrology (Carlton) 12:275–284
41. Poon TC, Chiu CH, Lai PB, Mok TS, Zee B, Chan AT, Sung JJ, Johnson PJ (2005) Correlation and prognostic significance of beta-galactoside alpha-2,6-sialyltransferase and serum monosialylated alpha-fetoprotein in hepatocellular carcinoma. World J Gastroenterol 11:6701–6706
42. Grunewald S, Matthijs G, Jaeken J (2002) Congenital disorders of glycosylation: a review. Pediatr Res 52:618–624
43. Stibler H, Allgulander C, Borg S, Kjellin KG (1978) Abnormal microheterogeneity of transferrin in serum and cerebrospinal fluid in alcoholism. Acta Med Scand 204:49–56
44. Ekuni A, Miyoshi E, Ko JH, Noda K, Kitada T, Ihara S, Endo T, Hino A, Honke K, Taniguchi N (2002) A glycomic approach to hepatic tumors in N-acetylglucosaminyltransferase III (GnT-III) transgenic mice induced by diethylnitrosamine (DEN): identification of haptoglobin as a target molecule of GnT-III. Free Radic Res 36:827–833
45. Li D, Mallory T, Satomura S (2001) AFP-L3: a new generation of tumor marker for hepatocellular carcinoma. Clin Chim Acta 313:15–19
46. Poon TC, Mok TS, Chan AT, Chan CM, Leong V, Tsui SH, Leung TW, Wong HT, Ho SK, Johnson PJ (2002) Quantification and utility of monosialylated alpha-fetoprotein in the diagnosis of hepatocellular carcinoma with nondiagnostic serum total alpha-fetoprotein. Clin Chem 48:1021–1027
47. Noda K, Miyoshi E, Uozumi N, Yanagidani S, Ikeda Y, Gao C, Suzuki K, Yoshihara H, Yoshikawa K, Kawano K, Hayashi N, Hori M, Taniguchi N (1998) Gene expression of alpha1-6 fucosyltransferase in human hepatoma tissues: a possible implication for increased fucosylation of alpha-fetoprotein. Hepatology 28:944–952
48. Wada Y, Azadi P, Costello CE, Dell A, Dwek RA, Geyer H, Geyer R, Kakehi K, Karlsson NG, Kato K, Kawasaki N, Khoo KH, Kim S, Kondo A, Lattova E, Mechref Y, Miyoshi E, Nakamura K, Narimatsu H, Novotny MV, Packer NH, Perreault H, Peter-Katalinic J, Pohlentz G, Reinhold VN, Rudd PM, Suzuki A, Taniguchi N (2007) Comparison of the methods for profiling glycoprotein glycans–HUPO human disease glycomics/proteome initiative multi-institutional study. Glycobiology 17:411–422
49. Wada Y, Gu J, Okamoto N, Inui K (1994) Diagnosis of carbohydrate-deficient glycoprotein syndrome by matrix-assisted laser desorption time-of-flight mass spectrometry. Biol Mass Spectrom 23:108–109
50. Wada Y (2006) Mass spectrometry for congenital disorders of glycosylation, CDG. J Chromatogr B Analyt Technol Biomed Life Sci 838:3–8
51. An HJ, Miyamoto S, Lancaster KS, Kirmiz C, Li B, Lam KS, Leiserowitz GS, Lebrilla CB (2006) Profiling of glycans in serum for the discovery of potential biomarkers for ovarian cancer. J Proteome Res 5:1626–1635
52. Kirmiz C, Li B, An HJ, Clowers BH, Chew HK, Lam KS, Ferrige A, Alecio R, Borowsky AD, Sulaimon S, Lebrilla CB, Miyamoto S (2007) A serum glycomics approach to breast cancer biomarkers. Mol Cell Proteomics 6:43–55
53. Kyselova Z, Mechref Y, Al Bataineh MM, Dobrolecki LE, Hickey RJ, Vinson J, Sweeney CJ, Novotny MV (2007) Alterations in the serum glycome due to metastatic prostate cancer. J Proteome Res 6:1822–1832

54. Goldman R, Ressom HW, Varghese RS, Goldman L, Bascug G, Loffredo CA, Abdel-Hamid M, Gouda I, Ezzat S, Kyselova Z, Mechref Y, Novotny MV (2009) Detection of hepatocellular carcinoma using glycomic analysis. Clin Cancer Res 15:1808–1813
55. Kyselova Z, Mechref Y, Kang P, Goetz JA, Dobrolecki LE, Sledge GW, Schnaper L, Hickey RJ, Malkas LH, Novotny MV (2008) Breast cancer diagnosis and prognosis through quantitative measurements of serum glycan profiles. Clin Chem 54:1166–1175
56. Mechref Y, Hussein A, Bekesova S, Pungpapong V, Zhang M, Dobrolecki LE, Hickey RJ, Hammoud ZT, Novotny MV (2009) Quantitative serum glycomics of esophageal adenocarcinoma and other esophageal disease onsets. J Proteome Res 8:2656–2666
57. Alley WR Jr, Vasseur JA, Goetz JA, Svoboda M, Mann BF, Matei DE, Menning N, Hussein A, Mechref Y, Novotny MV (2012) N-Linked glycan structures and their expressions change in the blood sera of ovarian cancer patients. J Proteome Res 11:2282–2300
58. Matsumoto K, Shimizu C, Arao T, Andoh M, Katsumata N, Kohno T, Yonemori K, Koizumi F, Yokote H, Aogi K, Tamura K, Nishio K, Fujiwara Y (2009) Identification of predictive biomarkers for response to trastuzumab using plasma FUCA activity and N-glycan identified by MALDI-TOF-MS. J Proteome Res 8:457–462
59. Nordhoff E, Ingendoh A, Cramer R, Overberg A, Stahl B, Karas M, Hillenkamp F, Crain PF (1992) Matrix-assisted laser desorption/ionization mass spectrometry of nucleic acids with wavelengths in the ultraviolet and infrared. Rapid Commun Mass Spectrom 6:771–776
60. Fu DJ, Broude NE, Köster H, Smith CL, Cantor CR (1996) Efficient preparation of short DNA sequence ladders potentially suitable for MALDI-TOF DNA sequencing. Genet Anal 12:137–142
61. Braun A, Little DP, Köster H (1997) Detecting CFTR gene mutations by using primer oligo base extension and mass spectrometry. Clin Chem 43:1151–1158
62. Ross P, Hall L, Smirnov I, Haff L (1998) High level multiplex genotyping by MALDI-TOF mass spectrometry. Nat Biotechnol 16:1347–1351
63. Luan J, Yuan J, Li X, Jin S, Yu L, Liao M, Zhang H, Xu C, He Q, Wen B, Zhong X, Chen X, Chan HL, Sung JJ, Zhou B, Ding C (2009) Multiplex detection of 60 hepatitis B virus variants by MALDI-TOF mass spectrometry. Clin Chem 55:1503–1509
64. Liao HK, Su YN, Kao HY, Hung CC, Wang HT, Chen YJ (2005) Parallel minisequencing followed by multiplex matrix-assisted laser desorption/ionization mass spectrometry assay for beta-thalassemia mutations. J Hum Genet 50:139–150
65. Farkas DH, Miltgen NE, Stoerker J, van den Boom D, Highsmith WE, Cagasan L, McCullough R, Mueller R, Tang L, Tynan J, Tate C, Bombard A (2010) The suitability of matrix assisted laser desorption/ionization time of flight mass spectrometry in a laboratory developed test using cystic fibrosis carrier screening as a model. J Mol Diagn 2:611–619
66. Ding C, Lo YM (2006) MALDI-TOF mass spectrometry for quantitative, specific, and sensitive analysis of DNA and RNA. Ann N Y Acad Sci 1075:282–287
67. Lo YM, Tsui NB, Chiu RW, Lau TK, Leung TN, Heung MM, Gerovassili A, Jin Y, Nicolaides KH, Cantor CR, Ding C (2007) Plasma placental RNA allelic ratio permits noninvasive prenatal chromosomal aneuploidy detection. Nat Med 13:218–223
68. Hidaka H, Hanyu N, Sugano M, Kawasaki K, Yamauchi K, Katsuyama T (2007) Analysis of human serum lipoprotein lipid composition using MALDI-TOF mass spectrometry. Ann Clin Lab Sci 37:213–221
69. Miura D, Fujimura Y, Tachibana H, Wariishi H (2010) Highly sensitive matrix-assisted laser desorption ionization-mass spectrometry for high-throughput metabolic profiling. Anal Chem 82:498–504
70. Vermillion-Salsbury RL, Hercules DM (2002) 9-Aminoacridine as a matrix for negative mode matrix-assisted laser desorption/ionization. Rapid Commun Mass Spectrom 16: 1575–1581
71. Shroff R, Muck A, Svatos A (2007) Analysis of low molecular weight acids by negative mode matrix-assisted laser desorption/ionization time-of-flight mass spectrometry. Rapid Commun Mass Spectrom 21:3295–3300

72. Mims D, Hercules D (2004) Quantification of bile acids directly from plasma by MALDI-TOF-MS. Anal Bioanal Chem 378:1322–1326
73. Law KP, Larkin JR (2011) Recent advances in SALDI-MS techniques and their chemical and bioanalytical applications. Anal Bioanal Chem 399:2597–2622
74. Sunner J, Dratz E, Chen YC (1995) Graphite surface-assisted laser desorption/ionization time-of-flight mass spectrometry of peptides and proteins from liquid solutions. Anal Chem 67:4335–4342
75. Go EP, Prenni JE, Wei J, Jones A, Hall SC, Witkowska HE, Shen Z, Siuzdak G (2003) Desorption/ionization on silicon time-of-flight/time-of-flight mass spectrometry. Anal Chem 75:2504–2506
76. Ren SF, Zhang L, Cheng ZH, Guo YL (2005) Immobilized carbon nanotubes as matrix for MALDI-TOF-MS analysis: applications to neutral small carbohydrates. J Am Soc Mass Spectrom 16:333–339
77. Zhang J, Wong HY, Guo YL (2005) Amino acids analysis by MALDI mass spectrometry using carbon nanotube as matrix. Chin J Chem 23:185–189
78. Dong X, Cheng J, Li J, Wang Y (2010) Graphene as a novel matrix for the analysis of small molecules by MALDI-TOF MS. Anal Chem 82:6208–6214
79. Zhou X, Wei Y, He Q, Boey F, Zhang Q, Zhang H (2010) Reduced graphene oxide films used as matrix of MALDI-TOF-MS for detection of octachlorodibenzo-p-dioxin. Chem Commun (Camb) 46:6974–6976
80. Soltzberg LJ, Patel P (2004) Small molecule matrix-assisted laser desorption/ionization time-of-flight mass spectrometry using a polymer matrix. Rapid Commun Mass Spectrom 18: 1455–1458
81. Su CL, Tseng WL (2007) Gold nanoparticles as assisted matrix for determining neutral small carbohydrates through laser desorption/ionization time-of-flight mass spectrometry. Anal Chem 79:1626–1633
82. Sanguinet L, Alévêque O, Blanchard P, Dias M, Levillain E, Rondeau D (2006) Desorption/ ionization on self-assembled monolayer surfaces (DIAMS). J Mass Spectrom 41:830–833
83. Lu M, Lai Y, Chen G, Cai Z (2011) Laser desorption/ionization on the layer of graphene nanoparticles coupled with mass spectrometry for characterization of polymers. Chem Commun (Camb) 47:12807–12809
84. Metz TO, Zhang Q, Page JS, Shen Y, Callister SJ, Jacobs JM, Smith RD (2007) The future of liquid chromatography-mass spectrometry (LC-MS) in metabolic profiling and metabolomic studies for biomarker discovery. Biomark Med 1:159–185
85. Edwards JL, Kennedy RT (2005) Metabolomic analysis of eukaryotic tissue and prokaryotes using negative mode MALDI time-of-flight mass spectrometry. Anal Chem 77:2201–2209
86. Vaidyanathan S, Goodacre R (2007) Quantitative detection of metabolites using matrix-assisted laser desorption/ionization mass spectrometry with 9-aminoacridine as the matrix. Rapid Commun Mass Spectrom 21:2072–2078
87. Gogichaeva NV, Alterman MA (2012) Amino acid analysis by means of MALDI TOF mass spectrometry or MALDI TOF/TOF tandem mass spectrometry. Methods Mol Biol 828: 121–135
88. Ng EW, Lam HS, Ng PC, Poon TC (2012) Quantification of citrulline by parallel fragmentation monitoring – a novel method using graphitized carbon nanoparticles and MALDI-TOF/ TOF mass spectrometry. Clin Chim Acta. doi:10.1016/j.cca.2012.10.039
89. Lowenthal MS, Yen J, Bunk DM, Phinney KW (2010) Certification of NIST standard reference material 2389a, amino acids in 0.1 mol/L HCl-quantification by ID LC-MS/MS. Anal Bioanal Chem 397:511–519
90. Shin S, Fung SM, Mohan S, Fung HL (2011) Simultaneous bioanalysis of L-arginine, L-citrulline, and dimethylarginines by LC-MS/MS. J Chromatogr B Analyt Technol Biomed Life Sci 879:467–474

91. Gobey J, Cole M, Janiszewski J, Covey T, Chau T, Kovarik P, Corr J (2005) Characterization and performance of MALDI on a triple quadrupole mass spectrometer for analysis and quantification of small molecules. Anal Chem 77:5643–5654
92. Volmer DA, Sleno L, Bateman K, Sturino C, Oballa R, Mauriala T, Corr J (2007) Comparison of MALDI to ESI on a triple quadrupole platform for pharmacokinetic analyses. Anal Chem 79:9000–9006
93. Meesters RJ, van Kampen JJ, Scheuer RD, van der Ende ME, Gruters RA, Luider TM (2011) Determination of the antiretroviral drug tenofovir in plasma from HIV-infected adults by ultrafast isotope dilution MALDI-triple quadrupole tandem mass spectrometry. J Mass Spectrom 46:282–289
94. van Kampen JJ, Reedijk ML, Burgers PC, Dekker LJ, Hartwig NG, van der Ende IE, de Groot R, Osterhaus AD, Burger DM, Luider TM, Gruters RA (2010) Ultra-fast analysis of plasma and intracellular levels of HIV protease inhibitors in children: a clinical application of MALDI mass spectrometry. PLoS One 5:e11409
95. Lesur A, Varesio E, Domon B, Hopfgartner G (2012) Peptides quantification by liquid chromatography with matrix-assisted laser desorption/ionization and selected reaction monitoring detection. J Proteome Res 11:4972–4982
96. Hutchens TW, Yip TT (1993) New desorption strategies for the mass spectrometric analysis of macromolecules. Rapid Commun Mass Spectrom 7:576–580
97. Poon TCW (2007) Opportunities and limitations of SELDI-TOF mass spectrometry in biomedical research – practical advices. Expert Rev Proteomics 4:51–65
98. Li J, Orlandi R, White CN, Rosenzweig J, Zhao J, Seregni E, Morelli D, Yu Y, Meng XY, Zhang Z, Davidson NE, Fung ET, Chan DW (2005) Independent validation of candidate breast cancer serum biomarkers identified by mass spectrometry. Clin Chem 51:2229–2235
99. Pang RT, Poon TC, Chan KC, Lee NL, Chiu RW, Tong YK, Wong RM, Chim SS, Ngai SM, Sung JJ, Lo YM (2006) Serum proteomic fingerprints of adult patients with severe acute respiratory syndrome. Clin Chem 52:421–429
100. Poon TC, Pang RT, Chan KC, Lee NL, Chiu RW, Tong YK, Chim SS, Ngai SM, Sung JJ, Lo YM (2012) Proteomic analysis reveals platelet factor 4 and beta-thromboglobulin as prognostic markers in severe acute respiratory syndrome. Electrophoresis 33:1894–1900
101. Roelofsen H, Alvarez-Llamas G, Schepers M, Landman K, Vonk RJ (2007) Proteomics profiling of urine with surface enhanced laser desorption/ionization time of flight mass spectrometry. Proteome Sci 5:2
102. Tang LA, Wang J, Loh KP (2010) Graphene-based SELDI probe with ultrahigh extraction and sensitivity for DNA oligomer. J Am Chem Soc 132:10976–10977
103. Li J, Zhang Z, Rosenzweig J, Wang YY, Chan DW (2002) Proteomics and bioinformatics approaches for identification of serum biomarkers to detect breast cancer. Clin Chem 48: 1296–1304
104. Ebert MP, Meuer J, Wiemer JC, Schulz HU, Reymond MA, Traugott U, Malfertheiner P, Röcken C (2004) Identification of gastric cancer patients by serum protein profiling. J Proteome Res 3:1261–1266
105. Semmes OJ, Feng Z, Adam BL, Banez LL, Bigbee WL, Campos D, Cazares LH, Chan DW, Grizzle WE, Izbicka E, Kagan J, Malik G, McLerran D, Moul JW, Partin A, Prasanna P, Rosenzweig J, Sokoll LJ, Srivastava S, Srivastava S, Thompson I, Welsh MJ, White N, Winget M, Yasui Y, Zhang Z, Zhu L (2005) Evaluation of serum protein profiling by surface-enhanced laser desorption/ionization time-of-flight mass spectrometry for the detection of prostate cancer: I. Assessment of platform reproducibility. Clin Chem 51:102–112
106. Poon TCW, Yip TT, Chan ATC, Yip C, Yip V, Mok TSK, Leung TWT, Ho S, Johnson PJ (2003) Comprehensive proteomic profiling identifies serum proteomic signatures for detection of hepatocellular carcinoma and its subtypes. Clin Chem 49:752–760
107. Forde CE, Gonzales AD, Smessaert JM, Murphy GA, Shields SJ, Fitch JP, McCutchen-Maloney SL (2002) A rapid method to capture and screen for transcription factors by SELDI mass spectrometry. Biochem Biophys Res Commun 290:1328–1335

108. Bane TK, LeBlanc JF, Lee TD, Riggs AD (2002) DNA affinity capture and protein profiling by SELDI-TOF mass spectrometry: effect of DNA methylation. Nucleic Acids Res 30:e69
109. Lehmann R, Melle C, Escher N, von Eggeling F (2005) Detection and identification of protein interactions of S100 proteins by ProteinChip technology. J Proteome Res 4: 1717–1721
110. Favre-Kontula L, Sattonnet-Roche P, Magnenat E, Proudfoot AE, Boschert U, Xenarios I, Vilbois F, Antonsson B (2008) Detection and identification of plasma proteins that bind GlialCAM using ProteinChip arrays, SELDI-TOF MS, and nano-LC MS/MS. Proteomics 8:378–388
111. Davies H, Lomas L, Austen B (1999) Profiling of amyloid beta peptide variants using SELDI protein chip arrays. Biotechniques 27:1258–1261
112. Tolson J, Bogumil R, Brunst E, Beck H, Elsner R, Humeny A, Kratzin H, Deeg M, Kuczyk M, Mueller GA, Mueller CA, Flad T (2004) Serum protein profiling by SELDI mass spectrometry: detection of multiple variants of serum amyloid alpha in renal cancer patients. Lab Invest 84:845–856
113. Schweigert FJ, Wirth K, Raila J (2004) Characterization of the microheterogeneity of transthyretin in plasma and urine using SELDI-TOF-MS immunoassay. Proteome Sci 2:5
114. Eriksson J, Woschnagg C, Fernvik E, Venge P (2007) A SELDI-TOF MS study of the genetic and post-translational molecular heterogeneity of eosinophil cationic protein. J Leukoc Biol 82:1491–1500
115. Poon TC, Chan KC, Ng PC, Chiu RW, Ang IL, Tong YK, Ng EK, Cheng FW, Li AM, Hon EK, Fok TF, Lo YM (2004) Serial analysis of plasma proteomic signatures in pediatric patients with severe acute respiratory syndrome and correlation with viral load. Clin Chem 50:1452–1455
116. Woodbury RL, McCarthy DL, Bulman AL (2012) Profiling of urine using ProteinChip® technology. Methods Mol Biol 818:97–107
117. Grus FH, Podust VN, Bruns K, Lackner K, Fu S, Dalmasso EA, Wirthlin A, Pfeiffer N (2005) SELDI-TOF-MS ProteinChip array profiling of tears from patients with dry eye. Invest Ophthalmol Vis Sci 46:863–876
118. Tomosugi N, Kitagawa K, Takahashi N, Sugai S, Ishikawa I (2005) Diagnostic potential of tear proteomic patterns in Sjögren's syndrome. J Proteome Res 4:820–825
119. Hida RY, Ohashi Y, Takano Y, Dogru M, Goto E, Fujishima H, Saito I, Saito K, Fukase Y, Tsubota K (2005) Elevated levels of human alpha-defensin in tears of patients with allergic conjunctival disease complicated by corneal lesions: detection by SELDI ProteinChip system and quantification. Curr Eye Res 30:723–730
120. Ranganathan S, Williams E, Ganchev P, Gopalakrishnan V, Lacomis D, Urbinelli L, Newhall K, Cudkowicz ME, Brown RH Jr, Bowser R (2005) Proteomic profiling of cerebrospinal fluid identifies biomarkers for amyotrophic lateral sclerosis. J Neurochem 95: 1461–1471
121. Simonsen AH, McGuire J, Podust VN, Davies H, Minthon L, Skoog I, Andreasen N, Wallin A, Waldemar G, Blennow K (2008) Identification of a novel panel of cerebrospinal fluid biomarkers for Alzheimer's disease. Neurobiol Aging 29:961–968
122. Siegmund R, Kiehntopf M, Deufel T (2009) Evaluation of two different albumin depletion strategies for improved analysis of human CSF by SELDI-TOF-MS. Clin Biochem 2: 1136–1143
123. Buhimschi CS, Bhandari V, Hamar BD, Bahtiyar MO, Zhao G, Sfakianaki AK, Pettker CM, Magloire L, Funai E, Norwitz ER, Paidas M, Copel JA, Weiner CP, Lockwood CJ, Buhimschi IA (2007) Proteomic profiling of the amniotic fluid to detect inflammation, infection, and neonatal sepsis. PLoS Med 4:e18
124. Park JS, Oh KJ, Norwitz ER, Han JS, Choi HJ, Seong HS, Kang YD, Park CW, Kim BJ, Jun JK, Syn HC (2008) Identification of proteomic biomarkers of preeclampsia in amniotic fluid using SELDI-TOF mass spectrometry. Reprod Sci 15:457–468

125. Ma Z, Liu C, Deng B, Dong S, Tao G, Zhan X, Wang C, Liu S, Qu X (2010) Different protein profile in amniotic fluid with nervous system malformations by surface-enhanced laser desorption-ionization/time-of-flight mass spectrometry (SELDI-TOF-MS) technology. J Obstet Gynaecol Res 36:1195–1203
126. Luciano-Montalvo C, Ciborowski P, Duan F, Gendelman HE, Meléndez LM (2008) Proteomic analyses associate cystatin B with restricted HIV-1 replication in placental macrophages. Placenta 29:1016–1023
127. Wibom C, Mörén L, Aarhus M, Knappskog PM, Lund-Johansen M, Antti H, Bergenheim AT (2009) Proteomic profiles differ between bone invasive and noninvasive benign meningiomas of fibrous and meningothelial subtype. J Neurooncol 94:321–331
128. Cadron I, Van Gorp T, Moerman P, Waelkens E, Vergote I (2011) Proteomic analysis of laser microdissected ovarian cancer tissue with SELDI-TOF MS. Methods Mol Biol 755:155–163
129. Poon TC, Sung JJ, Chow SM, Ng EK, Yu AC, Chu ES, Hui AM, Leung WK (2006) Diagnosis of gastric cancer by serum proteomic fingerprinting. Gastroenterology 130:1858–1864
130. Poon TC, Hui AY, Chan HL, Ang IL, Chow SM, Wong N, Sung JJ (2005) Prediction of liver fibrosis and cirrhosis in chronic hepatitis B infection by serum proteomic fingerprinting: a pilot study. Clin Chem 51:328–335
131. Liu W, Li X, Ding F, Li Y (2008) Using SELDI-TOF MS to identify serum biomarkers of rheumatoid arthritis. Scand J Rheumatol 37:94–102
132. Fung ET, Yip TT, Lomas L, Wang Z, Yip C, Meng XY, Lin S, Zhang F, Zhang Z, Chan DW, Weinberger SR (2005) Classification of cancer types by measuring variants of host response proteins using SELDI serum assays. Int J Cancer 115:783–789
133. Shi L, Zhang J, Wu P, Feng K, Li J, Xie Z, Xue P, Cai T, Cui Z, Chen X, Hou J, Zhang J, Yang F (2009) Discovery and identification of potential biomarkers of pediatric acute lymphoblastic leukemia. Proteome Sci 7:7
134. Sreseli RT, Binder H, Kuhn M, Digel W, Veelken H, Sienel W, Passlick B, Schumacher M, Martens UM, Zimmermann S (2010) Identification of a 17-protein signature in the serum of lung cancer patients. Oncol Rep 24:263–270
135. Ward DG, Wei W, Buckels J, Taha AM, Hegab B, Tariciotti L, Salih R, Qi YQ, Martin A, Johnson PJ (2010) Detection of pancreatic adenocarcinoma using circulating fragments of fibrinogen. Eur J Gastroenterol Hepatol 22:1358–1363
136. Ziegler ME, Chen T, LeBlanc JF, Wei X, Gjertson DW, Li KC, Khalighi MA, Lassman CR, Veale JL, Gritsch HA, Reed EF (2011) Apolipoprotein A1 and C-terminal fragment of α-1 antichymotrypsin are candidate plasma biomarkers associated with acute renal allograft rejection. Transplantation 92:388–395
137. Johnston O, Cassidy H, O'Connell S, O'Riordan A, Gallagher W, Maguire PB, Wynne K, Cagney G, Ryan MP, Conlon PJ, McMorrow T (2011) Identification of β2-microglobulin as a urinary biomarker for chronic allograft nephropathy using proteomic methods. Proteomics Clin Appl 5:422–431
138. Zhang Q, Wang J, Dong R, Yang S, Zheng S (2011) Identification of novel serum biomarkers in child nephroblastoma using proteomics technology. Mol Biol Rep 38:631–638
139. Flood-Nichols SK, Tinnemore D, Wingerd MA, Abu-Alya AI, Napolitano PG, Stallings JD, Ippolito DL (2012) Longitudinal analysis of maternal plasma apolipoproteins in pregnancy: a targeted proteomics approach. Mol Cell Proteomics. doi:10.1074/mcp.M112.018192
140. Poon TC, Johnson PJ (2001) Proteome analysis and its impact on the discovery of serological tumor markers. Clin Chim Acta 313:231–239
141. Ang IL, Poon TC, Lai PB, Chan AT, Ngai SM, Hui AY, Johnson PJ, Sung JJ (2006) Study of serum haptoglobin and its glycoforms in the diagnosis of hepatocellular carcinoma: a glycoproteomic approach. J Proteome Res 5:2691–2700
142. Jain MR, Bian S, Liu T, Hu J, Elkabes S, Li H (2009) Altered proteolytic events in experimental autoimmune encephalomyelitis discovered by iTRAQ shotgun proteomics analysis of spinal cord. Proteome Sci 7:25

143. Toyama A, Nakagawa H, Matsuda K, Ishikawa N, Kohno N, Daigo Y, Sato TA, Nakamura Y, Ueda K (2011) Deglycosylation and label-free quantitative LC-MALDI MS applied to efficient serum biomarker discovery of lung cancer. Proteome Sci 9:18
144. Li Y, Zhou K, Zhang Z, Sun L, Yang J, Zhang M, Ji B, Tang K, Wei Z, He G, Gao L, Yang L, Wang P, Yang P, Feng G, He L, Wan C (2012) Label-free quantitative proteomic analysis reveals dysfunction of complement pathway in peripheral blood of schizophrenia patients: evidence for the immune hypothesis of schizophrenia. Mol Biosyst 2012:2664–2671
145. Petricoin EF, Ardekani AM, Hitt BA, Levine PJ, Fusaro VA, Steinberg SM, Mills GB, Simone C, Fishman DA, Kohn EC, Liotta LA (2002) Use of proteomic patterns in serum to identify ovarian cancer. Lancet 359:572–577
146. Sorace JM, Zhan M (2003) A data review and re-assessment of ovarian cancer serum proteomic profiling. BMC Bioinformatics 4:24
147. Baggerly KA, Morris JS, Coombes KR (2004) Reproducibility of SELDI-TOF protein patterns in serum: comparing datasets from different experiments. Bioinformatics 20: 777–785
148. Baggerly KA, Morris JS, Edmonson SR, Coombes KR (2005) Signal in noise: Evaluating reported reproducibility of serum proteomic tests for ovarian cancer. J Natl Cancer Inst 97: 307–309
149. Adam BL, Qu Y, Davis JW, Ward MD, Clements MA, Cazares LH, Semmes OJ, Schellhammer PF, Yasui Y, Feng Z, Wright GL Jr (2002) Serum protein fingerprinting coupled with a pattern-matching algorithm distinguishes prostate cancer from benign prostate hyperplasia and healthy men. Cancer Res 62:3609–3614
150. McLerran D, Grizzle WE, Feng Z, Bigbee WL, Banez LL, Cazares LH, Chan DW, Diaz J, Izbicka E, Kagan J, Malehorn DE, Malik G, Oelschlager D, Partin A, Randolph T, Rosenzweig N, Srivastava S, Srivastava S, Thompson IM, Thornquist M, Troyer D, Yasui Y, Zhang Z, Zhu L, Semmes OJ (2008) Analytical validation of serum proteomic profiling for diagnosis of prostate cancer: sources of sample bias. Clin Chem 54:44–52
151. Yip TT, Chan JW, Cho WC, Yip TT, Wang Z, Kwan TL, Law SC, Tsang DN, Chan JK, Lee KC, Cheng WW, Ma VW, Yip C, Lim CK, Ngan RK, Au JS, Chan A, Lim WW, Ciphergen SARS Proteomics Study Group (2005) Protein chip array profiling analysis in patients with severe acute respiratory syndrome identified serum amyloid a protein as a biomarker potentially useful in monitoring the extent of pneumonia. Clin Chem 51:47–55
152. Kang X, Xu Y, Wu X, Liang Y, Wang C, Guo J, Wang Y, Chen M, Wu D, Wang Y, Bi S, Qiu Y, Lu P, Cheng J, Xiao B, Hu L, Gao X, Liu J, Wang Y, Song Y, Zhang L, Suo F, Chen T, Huang Z, Zhao Y, Lu H, Pan C, Tang H (2005) Proteomic fingerprints for potential application to early diagnosis of severe acute respiratory syndrome. Clin Chem 51:56–64
153. Pang RT, Poon TC, Chan KC, Lee NL, Chiu RW, Tong YK, Chim SS, Sung JJ, Lo YM (2006) Serum amyloid A is not useful in the diagnosis of severe acute respiratory syndrome. Clin Chem 52:1202–1204
154. McKernan M, McMillan DC, Anderson JR, Angerson WJ, Stuart RC (2008) The relationship between quality of life (EORTC QLQ-C30) and survival in patients with gastro-oesophageal cancer. Br J Cancer 98:888–893
155. El-Serag HB, Rudolph KL (2007) Hepatocellular carcinoma: epidemiology and molecular carcinogenesis. Gastroenterology 132:2557–2576
156. Aldington S, Harwood M, Cox B, Weatherall M, Beckert L, Hansell A, Pritchard A, Robinson G, Beasley R, Cannabis and Respiratory Disease Research Group (2008) Cannabis use and risk of lung cancer: a case–control study. Eur Respir J 31:280–286
157. Jorde R, Saleh F, Figenschau Y, Kamycheva E, Haug E, Sundsfjord J (2005) Serum parathyroid hormone (PTH) levels in smokers and non-smokers. The fifth Tromsø study. Eur J Endocrinol 152:39–45
158. Zhang Z, Bast RC Jr, Yu Y, Li J, Sokoll LJ, Rai AJ, Rosenzweig JM, Cameron B, Wang YY, Meng XY, Berchuck A, Van Haaften-Day C, Hacker NF, de Bruijn HW, van der Zee AG, Jacobs IJ, Fung ET, Chan DW (2004) Three biomarkers identified from serum proteomic analysis for the detection of early stage ovarian cancer. Cancer Res 64:5882–5890

159. Zhang Z, Chan DW (2005) Cancer proteomics: in pursuit of "true" biomarker discovery. Cancer Epidemiol Biomarkers Prev 14:2283–2286
160. Rai AJ, Zhang Z, Rosenzweig J, IeM S, Pham T, Fung ET, Sokoll LJ, Chan DW (2002) Proteomic approaches to tumor marker discovery. Arch Pathol Lab Med 126:1518–1526
161. Zhang Z, Chan DW (2010) The road from discovery to clinical diagnostics: lessons learned from the first FDA-cleared in vitro diagnostic multivariate index assay of proteomic biomarkers. Cancer Epidemiol Biomarkers Prev 19:2995–2999

Top Curr Chem (2014) 336: 177–196
DOI: 10.1007/128_2012_354

Published online: 22 August 2012

Application of Mass Spectrometry in Newborn Screening: About Both Small Molecular Diseases and Lysosomal Storage Diseases

Wuh-Liang Hwu, Yin-Hsiu Chien, Ni-Chung Lee, Shiao-Fang Wang, Shu-Chuan Chiang, and Li-Wen Hsu

Abstract Many genetic diseases, especially the inborn errors of metabolism, have very low incidences, so developing a newborn screening test for each disease is not practical. This obstacle was overcome by employing the tandem mass spectrometry (MS/MS) technology. In the analysis, the samples can be injected directly into the flowing system without passing through a column, and both acylcarnitine and amino acid profiles can be obtained at the same time. MS/MS newborn screening has been shown to improve the outcome of patients affected by a number of inborn errors of metabolism. Recently, MS/MS analytical methods were developed for second-tier tests of newborn screening; new substrates have also been developed to measure the activity of lysosomal enzymes so lysosomal storage diseases can be diagnosed by MS/MS method now.

Keywords Acylcarnitine · Amino acid · Lysosomal storage disease · Newborn screening · Second-tier · Tandem mass spectroscopy

Contents

1 Introduction to Newborn Screening 178
2 Development and Principle of MS/MS Newborn Screening 179
  2.1 Multiplexing: One Preparation for Many Molecules 179
  2.2 Principle of MS/MS 179
  2.3 Acylcarnitine and Amino Acid Profile Analysis 180
3 Impact of Tandem Mass Newborn Screening 181
  3.1 Incidence of Inborn Errors After MS/MS Screening 181
  3.2 Phenylketonuria 182
  3.3 Medium-Chain Acyl-CoA Dehydrogenase Deficiency 182
  3.4 Methylmalonic Acidemia 183

W.-L. Hwu (✉), Y.-H. Chien, N.-C. Lee, S.-F. Wang, S.-C. Chiang, and L.-W. Hsu
Department of Medical Genetics, National Taiwan University Hospital, Taipei, Taiwan
e-mail: hwuwlntu@ntu.edu.tw

3.5 Glutaric Aciduria Type I 184
3.6 Maple Syrup Urine Disease 185
3.7 Carnitine Uptake Defect 185
3.8 Ethical Consideration in MS/MS Screening 186
4 Second-Tier MS/MS Analysis 187
4.1 Principle of Second-Tier Test 187
4.2 Methylmalonic Acid for Methylmalonic Acidemia 187
4.3 Succinylacetone for Tyrosinemia 187
4.4 Congenital Adrenal Hyperplasia 188
5 MS/MS Screening for Lysosomal Storage Diseases (LSDs) 189
5.1 Introduction to Lysosomal Storage Diseases 189
5.2 Needs for Early Diagnosis of LSDs 189
5.3 Newborn Screening for LSDs 190
5.4 Measuring Enzyme Activity by MS/MS 190
5.5 Direct MS/MS Analysis of the Storage Material 191
6 Molecular Screening 192
References 192

1 Introduction to Newborn Screening

Newborn screening is a practice to test all newborn babies for diseases that would affect their health in the future [1]. The principle of newborn screening is to detect diseases before irreversible damages on tissue due to the screened diseases occur. Furthermore, through early institution of treatment, the prognosis of patients may also be improved. Historically, newborn screening started from the diagnosis of phenylketonuria and congenital hypothyroidism around 40 years ago. At that time, treatments for clinically diagnosed patients affected with these diseases were not successful. To perform newborn screening, blood from a heel prick was dried on a piece of filter paper. A simple and elegant assay, the bacterial inhibition assay, was first developed for phenylketonuria [2]. In the assay, a phenylalanine analog was added to the bacterial culture medium. Phenylalanine-dependent bacteria were then grown on the medium and only a dried blood spot (DBS) containing an excessive concentration of phenylalanine could support the growth of bacteria. The size of the growth zone was in proportion to the concentration of phenylalanine in the DBS. Bacterial inhibition assay was later applied to the analysis of other amino acids and galactose. Congenital hypothyroidism was tested initially by radioimmunoassay, and later replaced by immunoassays [3]. Congenital adrenal hyperplasia was also screened by immunoassay [4].

The criteria for newborn screening include that (1) the screening test must be simple and accurate, (2) patients must be picked up before disease onset, and (3) the treatment must be effective to prevent damage to the patients, etc. [5]. However, many genetic diseases, especially the inborn errors of metabolism, have very low incidences, so developing a test for each disease is not practical. This obstacle was finally overcome by the practice of newborn screening employing the tandem mass spectrometry (MS/MS) technology.

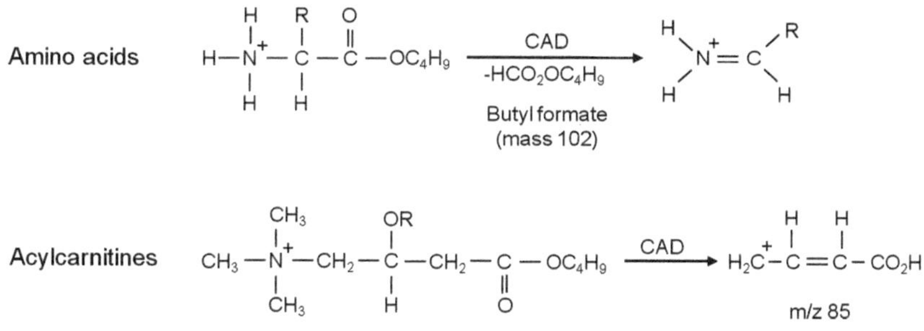

Fig. 1 Tandem mass spectrometry (MS/MS) analysis of amino acids and acylcarnitines. After collisionally activated dissociation (*CAD*), amino acids lose a neutral fragment of mass 102 (butyl formate) and acylcarnitines gives a positive ion of *m*/*z* (mass-to-charge) 85

2 Development and Principle of MS/MS Newborn Screening

2.1 Multiplexing: One Preparation for Many Molecules

Multiplexing has been the most critical concept and discovery for the advances of newborn screening in the past decade. Dr. Millington in the Duke University first developed the application of MS/MS in the diagnosis of fatty acid oxidation (FAO) defects [6]. In patients with FAO defects, fatty acyl-CoA of different lengths accumulated in the body. Carnitine is a carrier for fatty acid and carnitine binds to fatty acyl-CoA to form acylcarnitine. Carnitine can also bind to short chain acyl-CoA species that are generated from the metabolism of organic acids. Measurement of accumulation (elevated concentration) of acylcarnitines could therefore facilitate diagnosis of many FAO defects and organic acidemia. Similarly, amino acid profile analysis facilitates diagnosis of phenylketonuria and maple syrup urine disease [7, 8].

2.2 Principle of MS/MS

In MS/MS analysis the first mass (MS1) allows selected ions to enter the second mass (MS2) to be fragmented [9] and the third mass (MS3) then measure these fragments. Through computing, MS/MS can perform profile analysis in two modes including precursor ion scan and neutral loss scan (Fig. 1). When a group of ions generates the same daughter ion after fragmentation, a precursor ion scan can be used to recognize these ions all together. During a precursor ion scan, MS1 performs a scan function allowing ions with different *m*/*z* (mass-to-charge) ratios to enter MS2 to be fragmented; MS3 then monitors the appearance of the common daughter ion to recognize the precursors. When a group of ions generate a neutral fragment after fragmentation, then neutral loss scan can be used such that MS1 and MS3 perform parallel scans and MS3 monitors ions smaller than those measured in MS1 because of loss of the neutral fragment.

The achievement of high-throughput sample analysis – which is necessary for newborn screening – came with the application of electrospray ionization (ESI), in which a continuous flowing system could be sustained and the sample was injected into this flowing system [10]. Esterification of the acylcarnitines (most commonly as butyl ester) facilitates ionization and improves sensitivity. As ionization methods improved, e.g., through advances in ESI, underivatized samples could also be analyzed for many acylcarnitines and amino acids. Generally, however, in an identical analysis, sensitivity is less for underivatized acylcarnitines than for butyl esters [11].

2.3 Acylcarnitine and Amino Acid Profile Analysis

Acylcarnitine and amino acid analyses employ stable isotope-labeled internal standards to provide quantitative data [12]. Sample extract is evaporated under nitrogen gas and the residue is derivatized with *n*-butanolic HCl, changing the acylcarnitines to their *n*-butyl-esters. Chromatography is not usually necessary for ESI-MS/MS newborn screening, so the samples can be injected directly into this flowing system without passing through a column. When samples are analyzed by MS/MS, acylcarnitine species are identified by precursor ion scan through identifying ions which generate an 85-Da mass fragment (Fig. 2). This fragment ion is essentially the carnitine backbone produced by losses of butyl ester, quaternary ammonium, and the fatty acids. It is the reason that this fragment is common to both derivatized and underivatized fatty acylcarnitines of various chain lengths [11]. Quantitation is achieved by the comparison of the abundance of the various acylcarnitine species with the intensity of their closest isotope-labeled internal standard [13].

In the analysis of an amino acid profile, only α-amino acids fragment produces the neutral butyl formate molecule (102 Da). Therefore not all amino acids can be measured in a quantitative and specific manner by neutral loss scan. Leucine and isoleucine have the same mass so they appear as a single peak. Another two amino acid, cysteine and homocysteine, are not included in the routine neutral loss 102 panel in newborn screening because these compounds form disulfides and produce two sites of protonation and hence a doubly charged molecule (Fig. 3). Amino acids with a basic functional group also need to be analyzed with different scan functions. These amino acids include arginine, ornithine, and citrulline. Because these amino acids have a basic functional group that fragments easily, the most common fragments are a combination of butyl formate plus ammonia or basic groups. Hence, for ornithine and citrulline, a neutral loss of 119 Da scan is used [11].

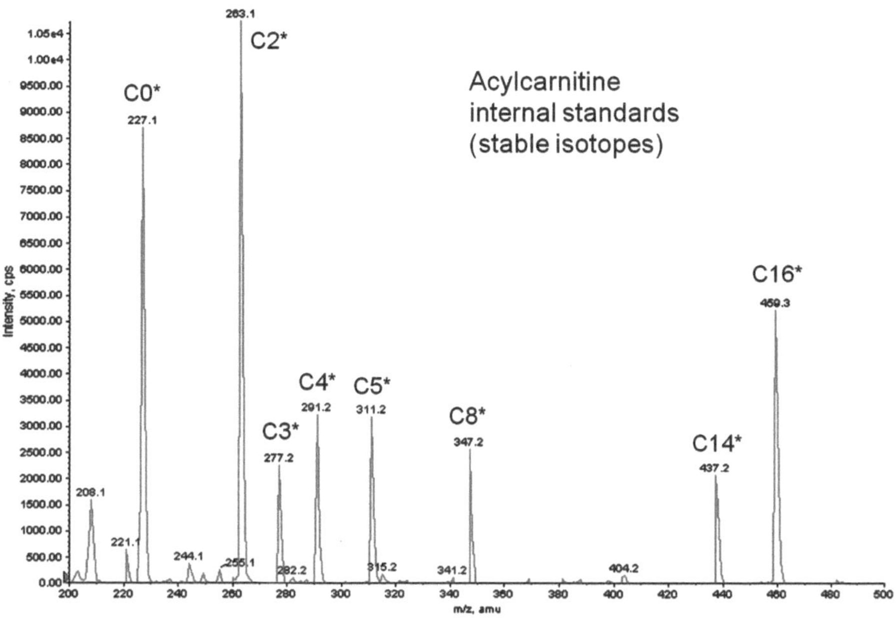

Fig. 2 Acylcarnitine profile. Chromatogram of a mixture of stable isotope internal standards. Either free carnitine (*C0**) or acylcarnitines with a chain-length of 2 to 16 carbons (C2*–C16*) were shown

3 Impact of Tandem Mass Newborn Screening

3.1 Incidence of Inborn Errors After MS/MS Screening

The recent development of ESI-MS/MS makes it possible to screen newborns for many rare inborn errors of metabolism. In one large-scale MS/MS screening program from Australia involving 362,000 newborns, the results revealed an elevated incidence of inborn errors. In the cohort screened with MS/MS, the prevalence of inborn errors, excluding phenylketonuria, was 15.7 per 100,000 births, as compared with adjusted rates of 8.6–9.5 per 100,000 births before the MS/MS screening [14]. In a follow-up report, MS/MS screening provides a better outcome for patients at 6 years of age, with fewer deaths and fewer clinically significant disabilities [15]. In a recent calculation, the incidence of FAO defects calculated from reports from Australia, Germany, and the USA involving a total of 5,256,999 newborns gives a combined incidence of all FAO defects of approximately 1 per 9,300 [16].

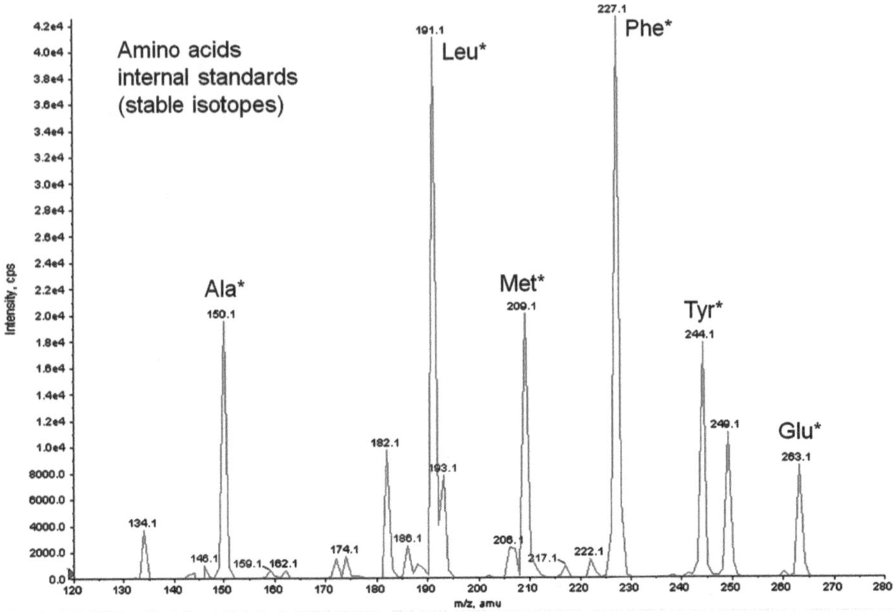

Fig. 3 Amino acid profile. Chromatogram of a mixture of stable isotope internal standards. Several amino acid internal standards including alanine (*Ala**), leucine (*Leu**), methionine (*Met**), tyrosine (*Tyr**), and glutamic acid (*Glu**) are labeled on the graph

3.2 Phenylketonuria

PKU is one of the most commonly screened diseases in all countries. After the institution of MS/MS, PKU is now screened by MS/MS (Fig. 4). Because of the high accuracy of MS/MS analysis of phenylalanine, the false-positive rate of PKU screening is reduced, and the cut-off value can be more precisely defined [17].

3.3 Medium-Chain Acyl-CoA Dehydrogenase Deficiency

A validated method for newborn screening for medium-chain acyl-CoA dehydrogenase (MCAD) deficiency was first published in 1993 and involved the detection of an elevation of octanoylcarnitine (C8-carnitine) [18]. MCAD deficiency is the commonest disorder of inborn errors of FAO with an incidence up to 1 per 10,000 in certain ethnic groups, due to a common c.985A>G mutation in the MCAD gene [19]. Children with MCAD deficiency may present acutely with hypoglycemia or a Reye-like illness, with a significant mortality and residual neurodevelopmental problems. In newborns with a positive acylcarnitine profile detected by newborn screening, the frequency of the c.985A>G mutant allele is lower than that observed

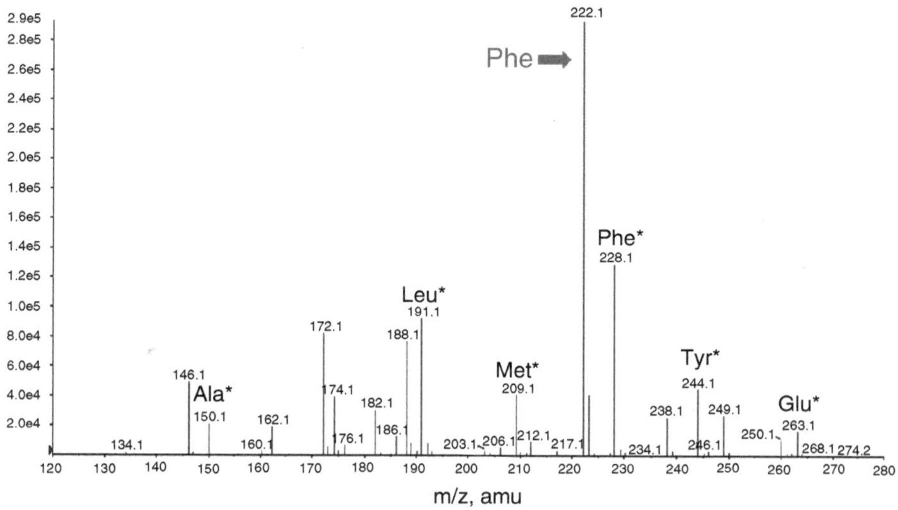

Fig. 4 An amino acid profile from a patient affected with phenylketonuria. The *arrow* points to an elevated peak of phenylalanine (*Phe*)

in clinically affected patients [20]. In Australia, MCAD-deficiency was 2.28 per 100,000 in the unscreened population and 5.2 per 100,000 in the screened population, and early diagnosis by screening reduced deaths and severe adverse events [15, 21]. MS/MS screening also detected women with MCAD deficiency following the identification of low free carnitine levels in their newborns [22].

3.4 *Methylmalonic Acidemia*

The methodology for the quantitative MS/MS analysis of propionylcarnitine (C3-carnitine), which is a marker for methylmalonic acidemia, in filter-paper blood specimens obtained from newborns was first reported in 2001 [23] (Fig. 5). Previously, methylmalonic acidemia was not included in newborn screening, and clinical suspicion depends on the manifestations of nausea, vomiting, consciousness disturbance, metabolic acidosis, and hyperammonemia. However, for the early-onset form of the disease, the first attack carries high mortality and morbidity rates. For the milder or later-onset form, brain damage still occurs because of delay in diagnosis. Now newborn screening provides a chance for early diagnosis of methylmalonic acidemia. Even though patients may already have symptoms before the result of screening can be available, those symptoms are nonspecific and information from newborn screening is always helpful. Decreased early mortality, less severe symptoms at diagnosis, and more favorable short-term neurodevelopmental outcomes have been reported in patients identified through MS/MS

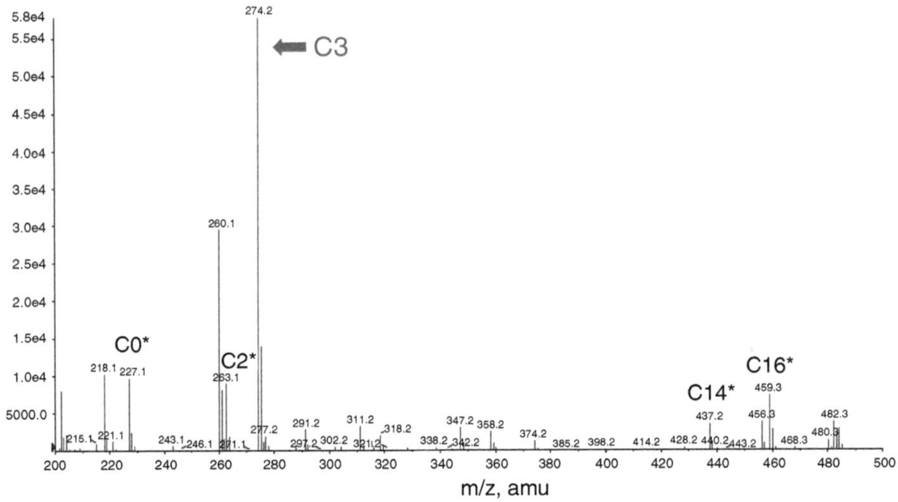

Fig. 5 An acylcarnitine profile from a patient affected with methylmalonic acidemia. The *arrow* points to an elevated peak of C3-carnitine (*C3*)

newborn screening [24]. However, the mild hyperammonemia that usually occurs before diagnosis still causes mild brain damage [25].

The false-positive rate is high in screening for methylmalonic acidemia using C3-carnitine as the biochemical marker. The high false-positive rate is a significant hindrance for rapid referral and aggressive treatment of babies suspected to have methylmalonic acidemia. Rapid second-tier testing is made possible by measuring DBS 3-hydroxy-propionic and methylmalonic acids levels [26] which will be discussed later.

3.5 Glutaric Aciduria Type I

Glutaric aciduria type I can be screened by the elevation of glutarylcarnitine (C5DC-carnitine) in the acylcarnitine profile. Before screening, patients with glutaric aciduria type I were recognized by clinical manifestations of hypotonia or spasticity [27]. Medical treatment is often not successful in those patients. When patients are detected by newborn screening, a low-lysine, low-tryptophan diet can be instituted immediately, and stress can be avoided to decrease the risk of encephalopathic crisis. Reports from several screening programs have demonstrated a significant improvement in outcome in the screened patients compared to the clinically diagnosed patients [28, 29]. However, according to current experience, false-negative screening, that is, newborns with the disease but no elevation of C5DC-carnitine, can also occur [30].

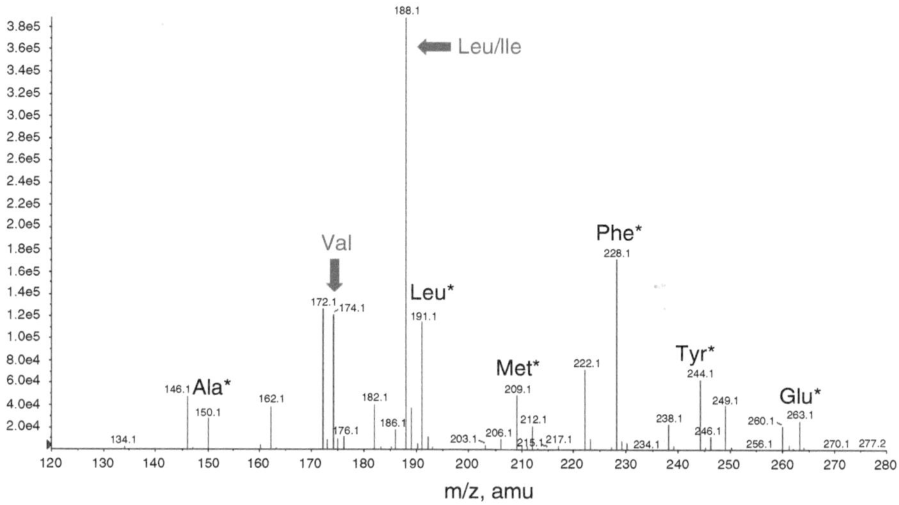

Fig. 6 An amino acid profile from a patient affected with maple syrup urine disease. The *arrows* point to elevated peaks of valine (*Val*) and leucine/isoleucine (*Leu/Ile*)

3.6 Maple Syrup Urine Disease

Clinical diagnosis of maple syrup urine disease (MSUD) is difficult because the symptoms, including intermittent consciousness disturbance and seizure, are not specific [31]. MSUD can be detected in newborn screening by the elevation of leucine/isoleucine, and valine [7] (Fig. 6). Leucine and isoleucine are not separated in the amino acid profile because there is no chromatographic separation prior to the newborn screening MS/MS analysis. MS/MS screening can make early diagnosis of MSUD and give a chance for early treatment [32, 33]. However, data from the California Newborn Screening Program revealed three missed cases of late-onset MSUD; second-tier testing with allo-isoleucine may improve sensitivity, but some children with variant forms will invariably be missed [33].

3.7 Carnitine Uptake Defect

Carnitine uptake defect (CUD), also known as primary carnitine deficiency, is an autosomal recessive disorder of FAO caused by mutations of the *SLC22A5* gene [34]. Carnitine is responsible for transporting fatty acids into mitochondria. Defective carnitine uptake results in urinary carnitine wasting and systemic and intracellular carnitine deficiencies [35]. Patients with CUD usually have cardiomyopathy, muscle weakness, recurrent hypoketotic hypoglycemic coma, or Reye-like syndrome [36, 37]. The recent implementation of MS/MS spectrometry screening of newborns has been critical for diagnosing CUD [14, 38, 39] (Fig. 7). A fetus with CUD can have a

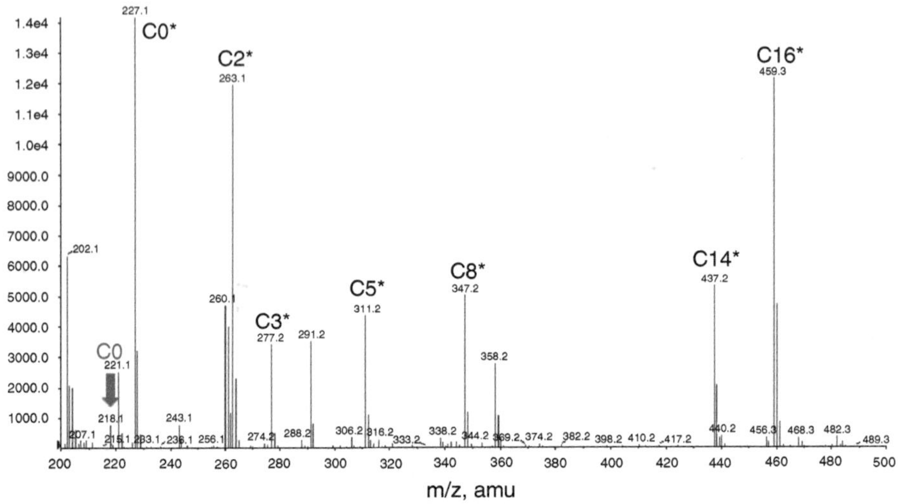

Fig. 7 An acylcarnitine profile from a patient affected with carnitine uptake defect. The levels of both free carnitine (*C0*, *arrow*) and acylcarnitines are decreased

carnitine supply from the mother. In addition, a normal fetus can have a low free carnitine level when the mother has carnitine deficiency. Diagnosis of mothers with CUD through newborn screening has also been demonstrated [35, 40].

3.8 Ethical Consideration in MS/MS Screening

MS/MS screening opens the possibility of finding diseases that are very rare. Because of the rarity of those diseases, their incidence, disease spectrum, natural history, and outcome of treatment may not be clear. According to previous criteria for institution of newborn screening [5], these diseases should not qualify for screening. However, one reason to have such strict criteria for institution of newborn screening previously was that screening by multiple assays led to a high cost. Now many diseases can be detected by a single analysis, and the shared cost for individual disease is very low. Currently most screening programs included 20–30 diseases in their MS/MS screening panel [41]. There are also several screening programs that are more conservative, like in the UK MS/MS analysis only generates data for MCAD deficiency [42]. In a recent report, accumulated evidence suggests all FAO defects should be included in MS/MS-based newborn screening programs provided that sufficient laboratory performance is guaranteed [16]. False-positives in MS/MS screening are certainly a concern. The infant who receive a false-positive result may be labeled as having a disease and therefore a cautious approach is needed [43].

4 Second-Tier MS/MS Analysis

4.1 *Principle of Second-Tier Test*

The acylcarnitine and amino acid profiles do not have high specificity or sensitivity for the screening of several particular diseases. Consequently the false-positive rates should be lowered at the expense of reduced sensitivity. It is possible that an additional biomarker can be analyzed from the original DBS to increase the specificity of the screening test, so most of the false-positive samples can be eliminated at this second-tier test stage without the need to recall the babies for a second DBS. A few successful examples are explained below.

4.2 *Methylmalonic Acid for Methylmalonic Acidemia*

The marker for the first-line screening of methylmalonic acidemia is C3-carnitine in the acylcarnitine profile. However, non-specific elevation of C3-carnitine can occur for many reasons including stress and prematurity, so the false-positive rate is high. Methylmalonic acid in DBS is a much more specific marker for methylmalonic acidemia. Methylmalonic acid can be extracted from the paper disk by methanol, separated with or without further derivation by a short chromatography, and then analyzed by MS/MS [23]. Therefore, direct analysis of methylmalonic acid from the same DBS that has elevated C3-carnitine level could be served as a second-tier test [26]. In several reports, this second-tier test significantly improved the performance of the screening for methylmalonic acidemia [26]. The analysis of methylmalonic acid is also helpful in monitoring the treatment of patients [44].

4.3 *Succinylacetone for Tyrosinemia*

Tyrosinemia type I or hepatorenal tyrosinemia is caused by a deficiency of fumarylacetoacetate hydrolase. Since this enzymatic step is not immediately after tyrosine, the concentration of tyrosine in affected newborns may overlap with normal newborns. Therefore, newborn screening for tyrosinemia by measuring DBS tyrosine level carries high false-positive and also high false-negative rates. Succinylacetone is a specific marker for tyrosinemia type I but it is not included in the amino acid or acylcarnitine profile. MS/MS analysis of succinylacetone can either serve as a second-tier test for the screening for tyrosinemia type I, or in areas of high prevalence of the disease the method used for the primary screening. Magera et al. described a precise measurement method for succinylacetone with the compound oximated, then extracted, butylated, and analyzed by liquid chromatography (LC)-MS/MS [45]. Measurement of succinylacetone has been effectively shown to pick up newborns affected by tyrosinemia type I [46].

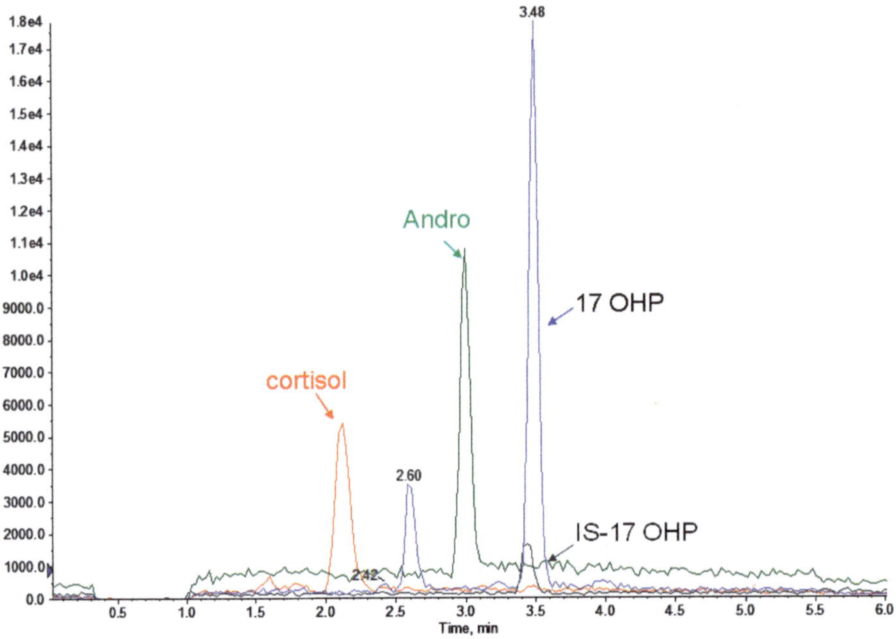

Fig. 8 LC-MS/MS analysis showing cortisol (*cortisol*), androstenedione (*andro*), and 17-hydroxy-progesterone (*17-OHP*). Because both cortisol and 17-OHP are both elevated, it is a false-positive sample due to stress response in newborns

4.4 Congenital Adrenal Hyperplasia

Congenital adrenal hyperplasia refers to a group of autosomal recessive, inherited disorders of cortisol biosynthesis. More than 90% of cases result from steroid 21-hydroxylase deficiency caused by mutations in *CYP21A2* gene [47]. This defect results in the deficiency of cortisol and the accumulation of 17-hydroxy-progesterone (17-OHP) and several other steroids. Severe disorders of corticosteroid deficiency will cause electrolyte imbalance, shock, or even death. In milder cases, the accumulated steroids induce the virilization of the external genital organs. Currently, newborn screening for congenital adrenal hyperplasia is conducted by measuring 17-OHP levels by immunoassay [48]. However, stress which occurred during the neonatal period induces the elevation of the steroids. In premature babies, high levels of fetal steroids also cross react with antibodies used in the immunoassay. MS/MS analysis of 17-OHP, together with cortisol and some other steroids, has been developed and served as a second-tier test for newborn screening for congenital adrenal hyperplasia [49, 50] (Fig. 8). These steroids can be extracted from DBS, separated by a short chromatography, and analyzed by MS/MS in several minutes. In normal newborns with other non-specific stress, elevation of 17-OHP is also accompanied by cortisol. Therefore, the second-tier test can reduced the false-positive rate.

5 MS/MS Screening for Lysosomal Storage Diseases (LSDs)

5.1 Introduction to Lysosomal Storage Diseases

Lysosomal Storage Diseases (LSDs) are a group of diseases caused by the dysfunction of lysosome, an organelle responsible for the degradation of intracellular waste materials. The most common cause of lysosomal dysfunction is the deficiency of one of the intra-lysosomal acid hydrolases [51]. Although individual lysosomal storage disorders are rare, their combined incidence has been estimated at 1 per 7,700 live births in Australia [52]. Clinical manifestations of LSDs depend on the type of materials stored in the lysosome and the type of cells or tissues involved. According to the storage materials, we can classify LSDs into mucopolysaccharidosis (MPS), oligosaccharidosis, lipidosis, and glycogenosis, etc. Depending on the tissues involved, patients can have clinical manifestations in the brain, eyes, skin, bones, and visceral organs, etc. That is, patients may have symptoms including mental retardation, cloudy cornea, thick skin and hair, short stature, bone and joint deformation, and hepatosplenomegaly, etc.

Laboratory diagnosis of LSDs can be achieved by analysis of the intracellular accumulated/stored material or the activity of the responsible enzyme. Analysis of the stored material often serves as a screening test for the presence of LSDs, for example, the dimethylmethylene blue test. Dimethylmethylene blue is a dye that binds to glycosaminoglycan, a compound excreted in elevated levels in the urine of patients affected by MPS. A positive, or abnormal, test result will then trigger the analysis of the activity of specific lysosomal enzymes to achieve specific diagnosis, for example, iduronidase for MPS I or iduronide sulfatase for MPS II. The principle of lysosomal enzyme activity measurement depends on the cleavage of a homologous substrate that has been labeled with either radioisotope or fluorescence tag. After the reaction, either the product of the reaction or the tag released from the reaction can be analyzed.

5.2 Needs for Early Diagnosis of LSDs

Specific treatments for LSDs have been developed recently [53]. The work started from purification of glucocerebrosidase from placenta to replace the enzyme deficient in Gaucher disease [54]. This method was later replaced by the production of enzymes by mammalian cells. The enzyme replacement therapy has progressed rapidly during the last 20 years. Currently recombinant enzymes are available for a panel of LSDs including Gaucher disease, Fabry disease [55, 56], Pompe disease [57, 58], MPS I [59], MPS II [60], and MPS VI [61]. However, although these drugs effectively remove the stored materials from the cells, residual disease sometimes significantly affects the outcome. For example, infants with Pompe disease have longer survival after enzyme therapy but many patients still need tracheostomy and mechanical ventilation. Residual bone disease is likely to be a general problem in the treatment of MPS by recombinant enzymes. Therefore, early diagnosis, best done by newborn screening, would be valuable in improving the outcome of LSD treatment [62].

5.3 *Newborn Screening for LSDs*

There have been a number of important technical developments that allow newborn screening for LSDs. For routine newborn screening, one 3.2-mm paper disc, containing around 3.2 μL of blood, punched from DBS is used for the assay. Chamoles et al. [63, 64] discovered that lysosomal enzymes on DBS were stable in enzyme assays and steady reaction rates were found during an extended incubation period of 20 h [63, 64]. Stability of enzymes during an extended incubation time allowed the development of assays using small sample volumes, and newborn screening for LSDs can be integrated into a standard screening menu.

Newborn screening for Pompe disease is the first example to demonstrate the value of newborn screening in LSDs. The Pompe disease newborn screening first developed in Taiwan employed a fluorescence artificial substrate [65] by modifying the Chamoles' DBS enzyme assay into a 96-well format. Acarbose was used as an inhibitor to prevent the interference of other acid glucosidase isoenzymes [66, 67]. There is now a digital microfluidic platform to perform rapid, multiplexed fluorometric enzymatic analysis of acid alpha-glucosidase and acid alpha-galactosidase to screen for Pompe and Fabry diseases [68].

5.4 *Measuring Enzyme Activity by MS/MS*

Measuring lysosomal enzyme activity by using fluorescence artificial substrate is a well-established methodology and has been available for diagnosis of most LSDs. However, when we want to do more than one enzyme tests at a time, these enzyme tests cannot be multiplexed. During the last few years, enzyme substrates in which the end products can be analyzed by MS/MS have been and are being developed. The first five-disease multiplex MS/MS enzyme substrate panel was developed by researchers in the University of Washington [66] (Fig. 9) and the reaction conditions were further optimized [69]. This panel includes Gaucher disease, Niemann-Pick A/B disease, Krabbe disease, Fabry disease, and Pompe disease. Li et al. attached a fatty acid that is variable in length to the substrates so that it would be easy to analyze several different substrate/product pairs at the same time. In the initial design, each reaction needed to be incubated individually, and the final product needed to be cleaned meticulously before MS/MS analysis. Recently, Spacil et al. modified their methods so three enzymes for Pompe disease, Fabry disease, and MPS I can be incubated together and the purification step was also

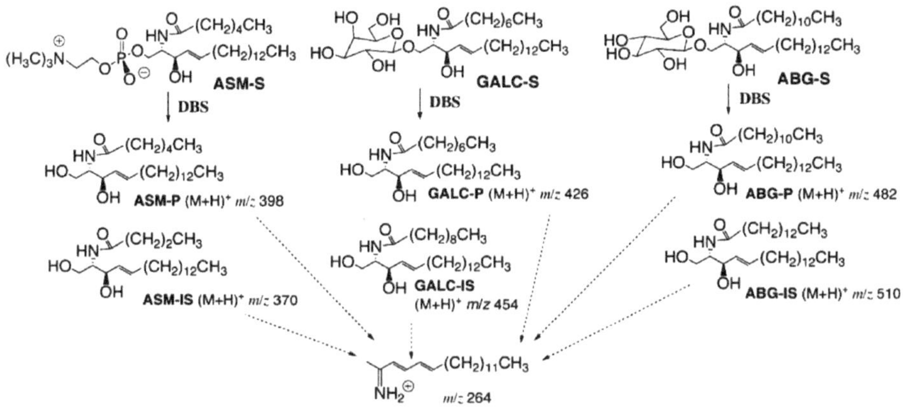

Fig. 9 Substrates (*S*), products (*P*), and internal standards (*IS*) for the three lysosomal enzyme assays. The masses of the products and internal standards are indicated. The ceramide products and internal standards of acid sphingomyelinase (*ASM*), galactocerebroside β-galactosidase (*GALC*), and acidβ-glucocerebrosidase (*ABG*) undergo collisionally activated dissociation to give the common imminium ion shown (*m/z* 264). GLA-P, GLA-IS, GAA-P, and GAA-IS undergo neutral-loss CID to give four different secondary ammonium ions. Enzymatic reactions with DBS are shown by *solid arrows* and CID by *dashed arrows* (from Li et al. [66])

simplified [70]. Currently, substrates for detection of MPS II, MPS IVA, and MPS VI have also been reported [71–73]. MS/MS DBS enzyme assay has just been tested for Gaucher disease, Pompe disease, Fabry disease, and Niemann-Pick disease type A/B in an anonymous prospective nationwide screening in Austria of more than 30,000 newborn babies. The results revealed a high overall incidence of 1 per 2,315 births among the Austrian population [74].

5.5 Direct MS/MS Analysis of the Storage Material

It will be much easier to screen for LSDs if there is a single biomarker for all or many LSDs. Glycosaminoglycan level rises in all MPS, and therefore analysis of glycosaminoglycan could be a valuable screening test. It has recently been published that dermatan sulfate and heparan sulfate might serve as a biomarker for newborn screening [75]. Disaccharides from DS, HS, and KS were digested by chondroitinase B, heparitinase, and keratanase and then analyzed by LC-MS/MS. Serum heparin cofactor II-thrombin complex (HCII-T), which is a glycosamino-glycan regulated serpin-protease complex, has recently been identified as a promising biomarker for both newborn screening and measurement of treatment outcomes in selected MPS diseases [76].

6 Molecular Screening

It is theoretically possible to detect all genetic diseases by molecular analysis, that is, by finding mutations in the genes. However, methods for both DNA extraction and mutation analysis are still tedious and large scale molecular screening is difficult. Newborn screening for cystic fibrosis employs a large amount of mutation analysis. The F508del mutation of the *CFTR* gene is distributed worldwide, reaching a carrier frequency of 80% in some Western European countries [77]. Molecular screening for cystic fibrosis is most often used as a second-tier test after the first screening by immunoreactive trypsinogen. In a proposed algorithm of cystic fibrosis screening, the second specimen is tested for 43 *CFTR* mutations [78].

Recently, molecular newborn screening has been used for the screening of severe combined immunodeficiency (SCID). SCID is a group of diseases caused by defects in the maturation of T lymphocyte. Deficiency in T cells leads to defects in both cellular and humoral immunity [79]. Patients usually die in infancy or childhood due to infection, but hematopoietic stem cell transplantation is a cure for SCID. Molecular screening for SCID elegantly targets on DNA structure originated from T cells. During T cell maturation, the T-cell receptor gene undergoes a series of recombination. A small circular DNA – the T-cell receptor excision circles (TRECs) – is then generated during the recombination process. SCID screening employs DBS DNA extraction and real time quantitative PCR [80]. Pilot SCID screening programs successfully detected newborns affected by SCID [81]. This screening test has been recommended in the US and all states need to accommodate this screening test according to their own timeline.

Acknowledgement and Declaration We thank our colleagues in the newborn screening laboratory for providing anonymous data for the chapter. There is no conflict of interest from any of the authors.

References

1. Kaye CI, Accurso F, La Franchi S, Lane PA, Northrup H, Pang S et al (2006) Introduction to the newborn screening fact sheets. Pediatrics 118(3):1304–1312
2. Guthrie R, Susi A (1963) A simple phenylalanine method for detecting phenylketonuria in large populations of newborn infants. Pediatrics 32:338–343
3. Dussault JH, Coulombe P, Laberge C, Letarte J, Guyda H, Khoury K (1975) Preliminary report on a mass screening program for neonatal hypothyroidism. J Pediatr 86(5):670–674
4. Pang S, Hotchkiss J, Drash AL, Levine LS, New MI (1977) Microfilter paper method for 17 alpha-hydroxyprogesterone radioimmunoassay: its application for rapid screening for congenital adrenal hyperplasia. J Clin Endocrinol Metab 45(5):1003–1008
5. Wilson JMG, Jungner G (1968) Principles and practice of screening for disease. HWO Chronicle 22(11):473
6. Millington DS, Terada N, Chace DH, Chen YT, Ding JH, Kodo N et al (1992) The role of tandem mass spectrometry in the diagnosis of fatty acid oxidation disorders. Prog Clin Biol Res 375:339–354

7. Chace DH, Hillman SL, Millington DS, Kahler SG, Roe CR, Naylor EW (1995) Rapid diagnosis of maple syrup urine disease in blood spots from newborns by tandem mass spectrometry. Clin Chem 41(1):62–68
8. Chace DH, Millington DS, Terada N, Kahler SG, Roe CR, Hofman LF (1993) Rapid diagnosis of phenylketonuria by quantitative analysis for phenylalanine and tyrosine in neonatal blood spots by tandem mass spectrometry. Clin Chem 39(1):66–71
9. Turecek F, Scott CR, Gelb MH (2007) Tandem mass spectrometry in the detection of inborn errors of metabolism for newborn screening. Methods Mol Biol 359:143–157
10. Rashed MS, Rahbeeni Z, Ozand PT (1999) Application of electrospray tandem mass spectrometry to neonatal screening. Semin Perinatol 23(2):183–193
11. Chace DH, Kalas TA, Naylor EW (2003) Use of tandem mass spectrometry for multianalyte screening of dried blood specimens from newborns. Clin Chem 49(11):1797–1817
12. Millington DS, Roe CR, Maltby DA (1984) Application of high resolution fast atom bombardment and constant B/E ratio linked scanning to the identification and analysis of acylcarnitines in metabolic disease. Biomed Mass Spectrom 11(5):236–241
13. Smith EH, Matern D (2010) Acylcarnitine analysis by tandem mass spectrometry. Curr Protoc Hum Genet; Chapter 17:Unit 17 8 1–20
14. Wilcken B, Wiley V, Hammond J, Carpenter K (2003) Screening newborns for inborn errors of metabolism by tandem mass spectrometry. N Engl J Med 348(23):2304–2312
15. Wilcken B, Haas M, Joy P, Wiley V, Bowling F, Carpenter K et al (2009) Expanded newborn screening: outcome in screened and unscreened patients at age 6 years. Pediatrics 124(2): e241–e248
16. Lindner M, Hoffmann GF, Matern D (2010) Newborn screening for disorders of fatty-acid oxidation: experience and recommendations from an expert meeting. J Inherit Metab Dis 33(5):521–526
17. Lukacs Z, Santer R (2006) Evaluation of electrospray-tandem mass spectrometry for the detection of phenylketonuria and other rare disorders. Mol Nutr Food Res 50(4–5):443–450
18. Chace DH, Hillman SL, Van Hove JL, Naylor EW (1997) Rapid diagnosis of MCAD deficiency: quantitative analysis of octanoylcarnitine and other acylcarnitines in newborn blood spots by tandem mass spectrometry. Clin Chem 43(11):2106–2113
19. Seddon HR, Gray G, Pollitt RJ, Iitia A, Green A (1997) Population screening for the common G985 mutation causing medium-chain acyl-CoA dehydrogenase deficiency with Eu-labeled oligonucleotides and the DELFIA system. Clin Chem 43(3):436–442
20. Andresen BS, Dobrowolski SF, O'Reilly L, Muenzer J, McCandless SE, Frazier DM et al (2001) Medium-chain acyl-CoA dehydrogenase (MCAD) mutations identified by MS/MS-based prospective screening of newborns differ from those observed in patients with clinical symptoms: identification and characterization of a new, prevalent mutation that results in mild MCAD deficiency. Am J Hum Genet 68(6):1408–1418
21. Wilcken B, Haas M, Joy P, Wiley V, Chaplin M, Black C et al (2007) Outcome of neonatal screening for medium-chain acyl-CoA dehydrogenase deficiency in Australia: a cohort study. Lancet 369(9555):37–42
22. Leydiker KB, Neidich JA, Lorey F, Barr EM, Puckett RL, Lobo RM et al (2011) Maternal medium-chain acyl-CoA dehydrogenase deficiency identified by newborn screening. Mol Genet Metab 103(1):92–95
23. Chace DH, DiPerna JC, Kalas TA, Johnson RW, Naylor EW (2001) Rapid diagnosis of methylmalonic and propionic acidemias: quantitative tandem mass spectrometric analysis of propionylcarnitine in filter-paper blood specimens obtained from newborns. Clin Chem 47(11):2040–2044
24. Dionisi-Vici C, Deodato F, Roschinger W, Rhead W, Wilcken B (2006) 'Classical' organic acidurias, propionic aciduria, methylmalonic aciduria and isovaleric aciduria: long-term outcome and effects of expanded newborn screening using tandem mass spectrometry. J Inherit Metab Dis 29(2–3):383–389

25. Lee NC, Chien YH, Peng SF, Huang AC, Liu TT, Wu AS et al (2008) Brain damage by mild metabolic derangements in methylmalonic acidemia. Pediatr Neurol 39(5):325–329
26. la Marca G, Malvagia S, Pasquini E, Innocenti M, Donati MA, Zammarchi E (2007) Rapid 2nd-tier test for measurement of 3-OH-propionic and methylmalonic acids on dried blood spots: reducing the false-positive rate for propionylcarnitine during expanded newborn screening by liquid chromatography-tandem mass spectrometry. Clin Chem 53(7):1364–1369
27. Strauss KA, Puffenberger EG, Robinson DL, Morton DH (2003) Type I glutaric aciduria, part 1: natural history of 77 patients. Am J Med Genet C Semin Med Genet 121C(1):38–52
28. Hsieh CT, Hwu WL, Huang YT, Huang AC, Wang SF, Hu MH et al (2008) Early detection of glutaric aciduria type I by newborn screening in Taiwan. J Formos Med Assoc 107(2):139–144
29. Boneh A, Beauchamp M, Humphrey M, Watkins J, Peters H, Yaplito-Lee J (2008) Newborn screening for glutaric aciduria type I in Victoria: treatment and outcome. Mol Genet Metab 94(3):287–291
30. Smith WE, Millington DS, Koeberl DD, Lesser PS (2001) Glutaric acidemia, type I, missed by newborn screening in an infant with dystonia following promethazine administration. Pediatrics 107(5):1184–1187
31. Ogier de Baulny H, Saudubray JM (2002) Branched-chain organic acidurias. Semin Neonatol 7(1):65–74
32. Huang HP, Chu KL, Chien YH, Wei ML, Wu ST, Wang SF et al (2006) Tandem mass neonatal screening in Taiwan – report from one center. J Formos Med Assoc 105(11):882–886
33. Puckett RL, Lorey F, Rinaldo P, Lipson MH, Matern D, Sowa ME et al (2010) Maple syrup urine disease: further evidence that newborn screening may fail to identify variant forms. Mol Genet Metab 100(2):136–142
34. Nezu J, Tamai I, Oku A, Ohashi R, Yabuuchi H, Hashimoto N et al (1999) Primary systemic carnitine deficiency is caused by mutations in a gene encoding sodium ion-dependent carnitine transporter. Nat Genet 21(1):91–94
35. Schimmenti LA, Crombez EA, Schwahn BC, Heese BA, Wood TC, Schroer RJ et al (2007) Expanded newborn screening identifies maternal primary carnitine deficiency. Mol Genet Metab 90(4):441–445
36. Tein I, De Vivo DC, Bierman F, Pulver P, De Meirleir LJ, Cvitanovic-Sojat L et al (1990) Impaired skin fibroblast carnitine uptake in primary systemic carnitine deficiency manifested by childhood carnitine-responsive cardiomyopathy. Pediatr Res 28(3):247–255
37. Stanley CA, DeLeeuw S, Coates PM, Vianey-Liaud C, Divry P, Bonnefont JP et al (1991) Chronic cardiomyopathy and weakness or acute coma in children with a defect in carnitine uptake. Ann Neurol 30(5):709–716
38. Schulze A, Lindner M, Kohlmuller D, Olgemoller K, Mayatepek E, Hoffmann GF (2003) Expanded newborn screening for inborn errors of metabolism by electrospray ionization-tandem mass spectrometry: results, outcome, and implications. Pediatrics 111(6 Pt 1):1399–1406
39. Wilcken B, Wiley V, Sim KG, Carpenter K (2001) Carnitine transporter defect diagnosed by newborn screening with electrospray tandem mass spectrometry. J Pediatr 138(4):581–584
40. Lee NC, Tang NL, Chien YH, Chen CA, Lin SJ, Chiu PC et al (2010) Diagnoses of newborns and mothers with carnitine uptake defects through newborn screening. Mol Genet Metab 100(1):46–50
41. Therrell BL, Adams J (2007) Newborn screening in North America. J Inherit Metab Dis 30(4):447–465
42. Downing M, Pollitt R (2008) Newborn bloodspot screening in the UK – past, present and future. Ann Clin Biochem 45(Pt 1):11–17
43. Tarini BA, Christakis DA, Welch HG (2006) State newborn screening in the tandem mass spectrometry era: more tests, more false-positive results. Pediatrics 118(2):448–456
44. Chen PW, Hwu WL, Ho MC, Lee NC, Chien YH, Ni YH et al (2010) Stabilization of blood methylmalonic acid level in methylmalonic acidemia after liver transplantation. Pediatr Transplant 14(3):337–341

45. Magera MJ, Gunawardena ND, Hahn SH, Tortorelli S, Mitchell GA, Goodman SI et al (2006) Quantitative determination of succinylacetone in dried blood spots for newborn screening of tyrosinemia type I. Mol Genet Metab 88(1):16–21
46. la Marca G, Malvagia S, Funghini S, Pasquini E, Moneti G, Guerrini R et al (2009) The successful inclusion of succinylacetone as a marker of tyrosinemia type I in Tuscany newborn screening program. Rapid Commun Mass Spectrom 23(23):3891–3893
47. Speiser PW, White PC (2003) Congenital adrenal hyperplasia. N Engl J Med 349(8):776–788
48. White PC (2009) Neonatal screening for congenital adrenal hyperplasia. Nat Rev Endocrinol 5(9):490–498
49. Lacey JM, Minutti CZ, Magera MJ, Tauscher AL, Casetta B, McCann M et al (2004) Improved specificity of newborn screening for congenital adrenal hyperplasia by second-tier steroid profiling using tandem mass spectrometry. Clin Chem 50(3):621–625
50. Minutti CZ, Lacey JM, Magera MJ, Hahn SH, McCann M, Schulze A et al (2004) Steroid profiling by tandem mass spectrometry improves the positive predictive value of newborn screening for congenital adrenal hyperplasia. J Clin Endocrinol Metab 89(8):3687–3693
51. Wenger DA, Coppola S, Liu SL (2003) Insights into the diagnosis and treatment of lysosomal storage diseases. Arch Neurol 60(3):322–328
52. Meikle PJ, Hopwood JJ, Clague AE, Carey WF (1999) Prevalence of lysosomal storage disorders. JAMA 281(3):249–254
53. Brady RO (2006) Enzyme replacement for lysosomal diseases. Annu Rev Med 57:283–296
54. Furbish FS, Blair HE, Shiloach J, Pentchev PG, Brady RO (1977) Enzyme replacement therapy in Gaucher's disease: large-scale purification of glucocerebrosidase suitable for human administration. Proc Natl Acad Sci USA 74(8):3560–3563
55. Schiffmann R, Murray GJ, Treco D, Daniel P, Sellos-Moura M, Myers M et al (2000) Infusion of alpha-galactosidase A reduces tissue globotriaosylceramide storage in patients with Fabry disease. Proc Natl Acad Sci USA 97(1):365–370
56. Eng CM, Guffon N, Wilcox WR, Germain DP, Lee P, Waldek S et al (2001) Safety and efficacy of recombinant human alpha-galactosidase A – replacement therapy in Fabry's disease. N Engl J Med 345(1):9–16
57. Kikuchi T, Yang HW, Pennybacker M, Ichihara N, Mizutani M, Van Hove JL et al (1998) Clinical and metabolic correction of pompe disease by enzyme therapy in acid maltase-deficient quail. J Clin Invest 101(4):827–833
58. Van den Hout JM, Reuser AJ, de Klerk JB, Arts WF, Smeitink JA, Van der Ploeg AT (2001) Enzyme therapy for pompe disease with recombinant human alpha-glucosidase from rabbit milk. J Inherit Metab Dis 24(2):266–274
59. Kakkis ED, Muenzer J, Tiller GE, Waber L, Belmont J, Passage M et al (2001) Enzyme-replacement therapy in mucopolysaccharidosis I. N Engl J Med 344(3):182–188
60. Muenzer J, Lamsa JC, Garcia A, Dacosta J, Garcia J, Treco DA (2002) Enzyme replacement therapy in mucopolysaccharidosis type II (Hunter syndrome): a preliminary report. Acta Paediatr Suppl 91(439):98–99
61. Harmatz P, Whitley CB, Waber L, Pais R, Steiner R, Plecko B et al (2004) Enzyme replacement therapy in mucopolysaccharidosis VI (Maroteaux–Lamy syndrome). J Pediatr 144(5):574–580
62. Kemper AR, Hwu WL, Lloyd-Puryear M, Kishnani PS (2007) Newborn screening for Pompe disease: synthesis of the evidence and development of screening recommendations. Pediatrics 120(5):e1327–e1334
63. Chamoles NA, Blanco M, Gaggioli D (2001) Fabry disease: enzymatic diagnosis in dried blood spots on filter paper. Clin Chim Acta 308(1–2):195–196
64. Chamoles NA, Blanco M, Gaggioli D (2001) Diagnosis of alpha-L-iduronidase deficiency in dried blood spots on filter paper: the possibility of newborn diagnosis. Clin Chem 47(4):780–781

65. Chien YH, Chiang SC, Zhang XK, Keutzer J, Lee NC, Huang AC et al (2008) Early detection of Pompe disease by newborn screening is feasible: results from the Taiwan screening program. Pediatrics 122(1):e39–e45
66. Li Y, Scott CR, Chamoles NA, Ghavami A, Pinto BM, Turecek F et al (2004) Direct multiplex assay of lysosomal enzymes in dried blood spots for newborn screening. Clin Chem 50(10):1785–1796
67. Zhang H, Kallwass H, Young SP, Carr C, Dai J, Kishnani PS et al (2006) Comparison of maltose and acarbose as inhibitors of maltase-glucoamylase activity in assaying acid alpha-glucosidase activity in dried blood spots for the diagnosis of infantile Pompe disease. Genet Med 8(5):302–306
68. Sista RS, Eckhardt AE, Wang T, Graham C, Rouse JL, Norton SM et al (2011) Digital microfluidic platform for multiplexing enzyme assays: implications for lysosomal storage disease screening in newborns. Clin Chem 57:1444–1451
69. Zhang XK, Elbin CS, Chuang WL, Cooper SK, Marashio CA, Beauregard C et al (2008) Multiplex enzyme assay screening of dried blood spots for lysosomal storage disorders by using tandem mass spectrometry. Clin Chem 54(10):1725–1728
70. Spacil Z, Elliott S, Reeber SL, Gelb MH, Scott CR, Turecek F (2011) Comparative triplex tandem mass spectrometry assays of lysosomal enzyme activities in dried blood spots using fast liquid chromatography: application to newborn screening of Pompe, Fabry, and Hurler diseases. Anal Chem 83(12):4822–4828
71. Duffey TA, Sadilek M, Scott CR, Turecek F, Gelb MH (2010) Tandem mass spectrometry for the direct assay of lysosomal enzymes in dried blood spots: application to screening newborns for mucopolysaccharidosis VI (Maroteaux–Lamy syndrome). Anal Chem 82(22):9587–9591
72. Khaliq T, Sadilek M, Scott CR, Turecek F, Gelb MH (2011) Tandem mass spectrometry for the direct assay of lysosomal enzymes in dried blood spots: application to screening newborns for mucopolysaccharidosis IVA. Clin Chem 57(1):128–131
73. Wolfe BJ, Blanchard S, Sadilek M, Scott CR, Turecek F, Gelb MH (2011) Tandem mass spectrometry for the direct assay of lysosomal enzymes in dried blood spots: application to screening newborns for mucopolysaccharidosis II (Hunter Syndrome). Anal Chem 83(3):1152–1156
74. Mechtler TP, Stary S, Metz TF, De Jesus VR, Greber-Platzer S, Pollak A et al (2012) Neonatal screening for lysosomal storage disorders: feasibility and incidence from a nationwide study in Austria. Lancet 379:335–341
75. Tomatsu S, Montano AM, Oguma T, Dung VC, Oikawa H, de Carvalho TG et al (2010) Dermatan sulfate and heparan sulfate as a biomarker for mucopolysaccharidosis I. J Inherit Metab Dis 33(2):141–150
76. Langford-Smith K, Arasaradnam M, Wraith JE, Wynn R, Bigger BW (2010) Evaluation of heparin cofactor II-thrombin complex as a biomarker on blood spots from mucopolysaccharidosis I, IIIA and IIIB mice. Mol Genet Metab 99(3):269–274
77. Worldwide survey of the delta F508 mutation – report from the cystic fibrosis genetic analysis consortium. Am J Hum Genet. 1990; 47(2):354–359
78. Sontag MK, Wright D, Beebe J, Accurso FJ, Sagel SD (2009) A new cystic fibrosis newborn screening algorithm: IRT/IRT1 upward arrow/DNA. J Pediatr 155(5):618–622
79. Buckley RH, Schiff RI, Schiff SE, Markert ML, Williams LW, Harville TO et al (1997) Human severe combined immunodeficiency: genetic, phenotypic, and functional diversity in one hundred eight infants. J Pediatr 130(3):378–387
80. Chan K, Puck JM (2005) Development of population-based newborn screening for severe combined immunodeficiency. J Allergy Clin Immunol 115(2):391–398
81. Verbsky JW, Baker MW, Grossman WJ, Hintermeyer M, Dasu T, Bonacci B et al (2012) Newborn screening for severe combined immunodeficiency; the Wisconsin experience (2008–2011). J Clin Immunol 32:82–88

Index

A
Achondroplasia, 49
Acid glucosidase, 190
Acid sphingomyelinase, 190
Acylcarnitines, 177–187
Acyl-CoA dehydrogenase deficiency, 182
Adrenal hyperplasia, congenital, 188
Albuminuria, MALDI-TOF, 148
Alpha-fetoprotein (AFP), 150
Amino acids, 68, 145, 177, 180, 187
  MALDI-TOF, 153
9-Aminoacridine, 153
Amyloid-beta, 148, 159
Aneuploidies, 33, 55, 61
  detection, 47, 55, 152
  fetal, 57–62
Array-based capture assays, 34
Aurora kinase, 89
Autosomal genes, 29
Autosomal disorders, 12
  dominant, 24–26, 49, 52
  recessive, 26, 30, 49, 52, 185

B
Binding site detection, 85
Biomarkers, 139
  quantification, 144
  verification, MRM, 128

C
Captopril, 83
Carnitine uptake defect (CUD), 185
Carrier testing, 31
Chaperones, 91
Charcot–Marie–Tooth disease, 32
Chondroitinase B, 191
Circulating fetal DNA, 47
Clinical diagnostics, 139
Clinical proteomics, 117
Concanavalin A, 146
Congenital adrenal hyperplasia, 188
Congenital disorders of glycosylation (CDG), 21, 27
  syndrome (CDGs), 149
Cyano-4-hydroxycinnamic acid, 142
Cyclin-dependent kinase 2, 89
Cyclophilin A, 88
Cysteine, 76
Cysteine protease inhibitors, 88
Cystic fibrosis, 152, 192
Cytokines, 111

D
Digital PCR, 99
Dihydroorotate dehydrogenase (DHODH), 31, 88
Dilated cardiomyopathy (DCM), 26
DNA, digital counting of cDNA, 99
  fetal, maternal plasma, 48
  plasma, 47
DNA polymerase, 3
DOPA responsive dystonia, 32
Down syndrome, 47, 55, 59
Droplet microfluidics, 99
Drug development, 67
  protein structure, 82
Druggability, 85
  index, 86

E

Enzyme activity, MS/MS, 190
Epidermal growth factor receptor (EGFR), 149
Epigenetic–genetic chromosome dosage, 58
Epigenetics, 12
  allelic ratio, 57
Exome sequencing, 9, 19, 31
Exopeptidase activity test, 147

F

Fabry disease, 189, 191
Fatty acids, 179, 185, 190
  oxidation (FAO) defects, 179
Fetal genome, genome-wide analysis, 61
Fibroblast growth factor receptor 2 (FGRF2), 78
First generation sequencing, 4
Fragment screening, 89
Freeman–Sheldon syndrome, 72
Fructose intolerance, 79
Fucosyltransferase, 150
Fumarate hydratase deficiency 80
Fumarylacetoacetate hydrolase, 187

G

G-protein coupled receptors (GPCR), 90
Gaucher disease, 189, 191
Gene enrichment methods, 24
Gene expression profile, 99
Genetic diseases, 67
Genome, 99
Genomic medicine, 1
Glutaric aciduria type I, 184
Glutarylcarnitine, 184
Glycine, 76
Glycoproteins, glycolipids/glycosylation, 149
Glycosaminoglycan, 189, 191
Glycosylation, 149

H

HeliScope sequencer, 5
Hemolysin, 9
Heparitinase, 191
Hepatocellular carcinoma (HCC), 150
Hereditary spastic paralegias (HSPs), 25
Heterogeneity, 99
Heteronuclear single-quantum correlation (HSQC), 69
High throughput screening (HTS), 83, 144
HLCS gene, 58
Homology modelling, 71
Hot spot mapping, 80
Human chorionic gonadotropin (hCG), 147
Human genetic diseases, 19
Human genome sequencing, 2
Human Genomic Project, 2
Huntington's disease, 49
17-Hydroxy-progesterone (17-OHP), 188
Hypertrophic cardiomyopathy, 21
Hypothyroidism, 178

I

Iduronide sulfatase, 189
Immunosorbent assay, 147, 160
Immunoturbidity assay, 147
Inflammatory bowel disease (IBD) 32
Integrated fluidic circuit chip (IFC chip) 106
Ion-sensitive field effect transistor (ISFET), 7

K

Kabuki syndrome, 31, 72
Keratinase, 191
Krabbe disease, 190

L

Ligand design, 88
Lipidosis, 189
Lysosomal hydrolases, 91
Lysosomal storage diseases (LSDs), 177, 189

M

Magnetic beads, 145
MALDI-MS, 130, 139
Maple syrup urine disease (MSUD), 185
Marfan syndrome (MFS), 21, 25
Massively parallel sequencing (MPS), 19
Matrix-assisted laser desorption/ionization (MALDI), 139
Matrix metalloproteinase, 89
Medium-chain acyl-CoA dehydrogenase (MCAD) deficiency, 182
Mendelian disorders, 72
Metabolites, quantification MALDI-TOF, 152
Methylation-sensitive restriction enzyme-mediated real-time PCR, 50
Methylmalonic acid, 187
Methylmalonic acidemia, 183
Microbial population analysis, 13
Microfluidics, 99
Miller syndrome, 31, 72
Misfolding, 67
  diseases, 75

Missense mutations, 67
  variations, 73
Mitochondrial disorders, 29
Mitochondrial genome, 29
Mitochondrial proteins, 29
Molecular diagnosis, 19
Molecular screening, 192
MS/MS, enzyme activity, 190
  screening, ethical considerations, 186
  second-tier, 187
Mucopolysaccharidosis (MPS), 189
Multi-gene disorders, 21
Multiple displacement amplification (MDA), 1ß3
Multiple reaction monitoring (MRM), 117
Multiplex, 117
  PCR, 33
Mutation, analysis, 67
  hot spots, 81
Mycobacterium tuberculosis, 85
Myosin, 79
Myotonic dystrophy, 49

N
Napthylphthalamic acid, 153
Neurofibromatosis type 1 (NF1), 21, 24
Newborn screening, 177, 181
Next generation sequencing (NGS), 1, 19, 71
  sequencer, 99
Niemann-Pick disease, 190
NMR, 69
Noninvasive prenatal diagnosis, 47

O
Octanoylcarnitine, 182
Oligonucleotide probe based capture, 34
Osteogenesis imperfecta, 78

P
Parkinson's disease, 76
Pathogen evolution, 13
Pathway analysis, MRM, 129
PCR, 33
  bias, 99
  digital, 53, 107
  drawbacks, 35
  multiplex, 33
  quantitative, 106
Peptide/protein biomarker discovery, functionalized magnetic beads, 145
Peripheral blood mononuclear cells (PBMCs), 32
Peroxisome proliferator-activated receptor, 89
Pharmacological chaperone therapy (PCT), 91
Phenylalanine hydroxylase (PAH), 75
Phenylketonuria, 75, 178
Plasma DNA, 47
Plasmodium falciparum, 88
Pompe disease, 189, 190
Post-genomics era, 67
Prenatal diagnosis, 33
  noninvasive, 47
Proline, 76, 153
Propionylcarnitine, 183
Protein–protein interactions, 79, 90
  disruption, 79
Protein Structure Database (PDB), 69
Protein structures, 67, 71
Proteomics, 117
Pyrosequencing, 5

Q
Quantitation, 117

R
Random sequencing analysis, 59
Relative chromosomal dosage (RCD) analysis, 57
Relative haplotype dosage (RHDO), 47, 61
Relative mutation dosage (RMD), 47, 52
Retinitis pigmentosa, 21, 27
Retinoschisin (RS), 78
Rhesus D, 49
RNA, allelic ratio, 55
  mRNA, 102
  sequencing (RNA-Seq), 12
Ryanodine receptors, 80

S
Sanger sequencing, 2, 33
Second generation sequencing, 5
Second-tier, 177
Selected reaction monitoring (SRM), 117
Sepiapterin reductase, 32
Sequence-specific exopeptidase activity test (SSEAT), 147
Sequencing-by-synthesis, 3
Serum amyloid A (SAA), 146
Serum heparin cofactor II-thrombin complex (HCII-T), 191
Serum transferrin, 149
Severe combined immunodeficiency (SCID), 192
Single cell analysis, 99
Single cell cDNA library, 99

Single-cell tagged reverse transcription (STRT), 108
Single nucleotide polymorphisms (SNPs), 11, 72, 152
SISCAPA (stable isotope standards and capture by anti-peptide antibodies), 130
Snyder–Robinson syndrome, 79
Spastic paraplegia, 21
Spermine synthase, 79
Steroid 21-hydroxylase deficiency, 188
Structural biology, 67
Structural genomics, 84
Structure-based drug design, 67, 84
Succinylacetone, 187
Surface-assisted laser desorption/ionization (SALDI), 139, 153
Surface-enhanced laser desorption/ionization (SELDI), 139

T
T-cell receptor excision circles (TRECs), 192
T-cell therapy, 111
Tandem mass spectroscopy, 177
Target-centric drug development, 82
Target gene enrichment, 19, 33
Targeted sequencing, 11
Tetrahydrobiopterin, 32
Thalassemia, 33, 62, 152
Third generation sequencing, 6
Time-of-flight (TOF), 139
Transcriptome, 12, 99
Tuberculosis structural genomics consortium, 85
Tunneling electron microscope (TEM), 8
Tyrosine, 32, 79, 182, 187
  kinase, 73, 78, 149
Tyrosinemia, 187

V
Virtual screening, 87

W
Whole exome sequencing (WES), 31
Whole genome sequencing (WGS), 11, 32

X
X-linked dyskeratosis congenita, 80
X-linked inhibitor of apoptosis (XIAP), 32
X-linked intellectual disability (XLID), 21, 28

Y
Y chromosome, 48

Z
Zero-mode wave guided (ZMW) nanostructures, 6

FSC
www.fsc.org
MIX
Papier | Fördert
gute Waldnutzung
FSC® C083411

Kolibri 360 GmbH
Ferdinand-Jühlke-Straße 7
99095 Erfurt, Deutschland
produktsicherheit@kolibri360.de